The
Productive
Electrician

2nd Edition

Michael Sammaritano

The Estimating Room Inc.

Library of Congress Cataloging-in-Publication Data
Michael Sammaritano
The Productive Electrician
Library of Congress Catalog Card No 2005937847
ISBN 0-9771541-1-4
List Price $49.50 USA Dollars
Printed in the United States of America

10 9 8 7 6 5 4 3 2
Visit our website at www.theestimatingroom.com

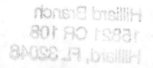

Introduction

To earn more, an electrician has to be more productive to himself and to his industry. Throughout this book, I refer to this individual as **The Productive Electrician.**

Dear Reader,

After some fifty years of exploring the ups and downs of Electrical Contracting, let me assure you that every method presented in this book has passed the test of time, and is brought to you in a unique platform in which a successful career can be built without the hassle and frustrations others have experienced.

I have written this book as if I were a reporter hired by each and every electrician who aspires to become a Productive Electrician. It is my intention to show you how to become a Productive Electrician and a high-wage earner, to educate those the industry penalizes for their lack of business skills, and to recruit the young people that every industry competes for.

For a fair assessment of your position in electrical contracting, you should ask yourself: "Did I choose electrical contracting solely for a paycheck or for the challenges it offers?"

If your choice is based on both, challenge and a good paycheck, bear in mind that most success stories are the product of people's commitment to their cause and personal sacrifice.

Electrical Contracting has built many successful careers; it also has made many individuals too complacent to work themselves into the ranks of Productive Electricians.

From my personal experience, I have concluded that what makes these individuals complacent, or, should I say, fearful of taking the next step is a lack of business skills, and therefore the confidence to overcome the obstacles each challenge presents.

You may ask "What do business skills have to do with an electrician who works for an electrical contractor?" and I'll say the difference between success and failure in your career. Stay the course and it will become apparent later on in this book.

Confidence, knowledge, and commitment are words that rarely amount to anything without the willingness to sacrifice. Of all the successful methods I have studied one stands out as supreme—the willingness to sacrifice for the sake of nourishing a rewarding career.

The sacrifice I refer to, as opposed to that of the fellow who just seeks wages, does not mean restraining your lifestyle to a poverty level. It simply means using your learning resources—time and money—judiciously.

In essence, if you want the kind of success you've seen others achieve, you must embrace commitment and sacrifice as part of your mission. Most of the methods presented in this book are meant to alleviate your degree of sacrifice but not your degree of commitment. In fact, you should intensify your commitment with each successful step you take. For example, have the commitment to learn as much as you can so you may contribute something of value to your job which, in turn, will pay you back many times over.

In life we rarely find those willing to tell us how things are done. And if and when we do, or discover the answers ourselves, it usually happens when we've lost the zeal to apply this newfound knowledge.

Best of luck,

—Michael Sammaritano

TABLE OF CONTENTS

Chapter 1

THE PRODUCTIVE ELECTRICIAN DEFINED

THE PRODUCTIVE ELECTRICIAN

For the purposes of this work, a Productive Electrician is defined as a person who qualifies as a Serviceman, Foreman, General-Foreman, or Project Manager for the electrical contracting marketplace.

The Productive Electrician, besides being a qualified tradesman, must possess basic business and technical skills that will help promote himself and his employer to the people in the industry they serve.

The status of Productive Electrician should not be confused with nor determined by the amount of conduit, wire, or the number of devices one is able to install in any given hour for that has nothing to do with the productive status one earns as a valued member of an electrical contractor's managerial team.

In that capacity, your productivity ratio is no longer measured by physical work output but by the effect your actions and decisions have on the company's bottom line. Therefore dedication, enthusiasm, and loyalty are your best tools.

THE JOB

The integrity of a Productive Electrician will not allow him to impose himself on a job; he will always earn his position.

In a workplace, you earn the position you're competing for by creating demand for your skills and personality. As a pro, you will always live up to, or exceed, your own expectations. When you make difficult performances look easy, people will cheer you the same way they cheer athletes. You'll hold the arena.

Along the way, you'll find that the best competitive edges are your positive attitude, your desire to learn, and your courage to apply the knowledge you possess in all you do. Developing this trio and taking them to higher levels will always create higher standards for your overall performance and build you a successful career supported by all those whom you work for.

For those who are new to this principle, be assured that it applies to all who are seeking a better job—beginners and veterans alike. Perhaps the following analogy on inflation and supply and demand will shed some light on the subject.

Inflation, to most people, is the difference between cost and value. If you pay a dollar for a service that is only worth seventy cents, then you lose thirty cents for every dollar you spend on that service—a thirty percent inflation, if you will. Conversely, if you pay a dollar for something that is worth one dollar and thirty cents, then your dollar has deflated by thirty percent—a thirty-cent gain in your dollar value. All things being equal, the latter will always translate into high demand and short supply assuring job security and financial success to all who help maintain that ratio.

This principle is mainly responsible for healthy relationships between employees and employers, the financial success or failure of individuals, companies, and nations. Even though inflation's inner workings are more complex—for there are many other contributing factors—productivity is the most important. Your degree of productivity will either increase or decrease the value of every dollar circulated—and most often affect the stability of the very job you hold.

Regardless of reasons or outcome, when we honestly produce to the best of our abilities, we've reached peak performance. We can average that performance into national statistics or we can individualize it by just concerning ourselves with our own.

In doing so, we can either be content and accept the results, or try to be more productive and seek ways and means to improve the value of our performance.

Individually, we can get to the core of the matter and conclude that no other formula than supply and demand can be more to our advantage. It rewards those who are willing to improve and penalizes those who are complacent and skeptical.

However, creating demand can be complex at times—especially when we cannot freely dedicate ourselves to that cause. Past experiences, greed, prejudices, and the like can interfere with our pride and natural desire to excel. Others' opinions can also inhibit our potential, thus affecting performance. For example, the traditional grandfathering of a job because someone in the family is a longtime employee of a company most often leads to complacency—an attitude responsible for massive layoffs.

A productive person must always pay attention to these factors. If not, he can easily get caught on the downswing and become a liability rather than a productive team member. This undoubtedly is a position you don't want to get caught in, for it will force you into imposing yourself on your job, rather than earning your position; thus decreasing self-esteem and job security substantially.

A productive person's job security lies in his ability to market his skills as a businessperson markets his products or services. As a free agent, he can contract himself for short or long periods with no fear of running out of work. For as long as he keeps himself updated on the latest methods and technology, throughout his career he may work for several employers. In fact, to stay in tune with the times, we may say he can schedule his career calendar with several contracts or on a job-by-job basis.

As an electrician, you are selling your talent to the company you work for every day. The "I'm making my boss rich." "Why should I produce more." or "The more I work, the less I get paid for" thoughts have no place in your formula for success. You are obliged to yourself and to your self-esteem to always be the best you can be in all you do. If you're productive and dedicated to the job, you can rest assured you will have a secure career with an income well above average.

Improving your overall performance and recognizing your potential can easily put you on top of the heap, creating high demand for your talent, which is your most precious asset. Your boss will recognize the increase in your value as a Productive Electrician long before you do. For it is in his best interest to spot and retain good, productive people.

THE ELECTRICIAN

An electrician is a person who installs, operates, maintains, or repairs electrical devices or electrical wiring. In addition, an electrician is also a person with feelings and dreams of his own.

If an electrician wants to excel beyond the point of just being a tradesman, such as a journeyman, serviceman, foreman, general foreman, or project manager, then to achieve his goal successfully he must expand his skills beyond technical workmanship—he must be proficient in the business side of contracting as well. He must also know how to deal with customers, manage employees, and get along with co-workers.

Typically, an electrician, as he comes up through the ranks, he will learn all about the rules and regulations of electrical installations, repairs, rules of good safety and workmanship. In contrast, he will not be taught basic business skills—which are essential to being a Productive Electrician.

Besides being a master of his trade, an electrician, without formal training, most often is expected to represent his employer as if he were a seasoned businessperson capable of selling, making business decisions, and executing complex contract documents. While this is ludicrous, unfortunately, it is very true in many small shops.

An electrician, due to his continuing exposure to intricate circuitry, complex job layouts, and electrical calculus, has developed a logical mind capable of interpreting and deciphering complex documents such as code books, plans, and specifications. Without any conscious effort on his part, this comes about by the time he rises to the level of mechanic.

If you have the desire to become a Productive Electrician, keep your dreams alive and declare yourself at the one-yard line of your aspiring career.

In this book, progressively and systematically, you will learn about marketing, contract interpretation, the change order and its relationship to the contract, proposal contracts, dealing with customers, the service call, and many other aspects of the electrical contracting that your job may call for.

Your representations and decisions will be logical and sound. Your newfound knowledge will give you the confidence to always be in control of your job and will keep you in high demand.

THE PRODUCTIVE ELECTRICIAN'S JOB

As a Productive Electrician, you must be able to switch jobs on demand. You may be needed as a serviceman on one occasion or as a foreman or even project manager on another. Regardless of which position you're called to do, be aware that everything you do or say is binding to the company you work for—your employer. At these times, you are the company.

The successful outcome of the business relationship between your employer and the customer is entirely up to you. Your attitude and your business and technical skills are direct reflections of yourself and of your employer.

In fact, more often than not, what attracts people to a company is one or more personal traits of the people who represent the company—you, rather than the company itself. People deal with people, not with objects or abstractions.

In short, the reason a Productive Electrician will always be in high demand is because he can easily fulfill these requirements successfully and productively—a win-win situation for the employer and the employee.

FINDING YOUR POSITION

Throughout your waking hours—whether you work for others, whether you want it or not—you are always marketing yourself and the company you work for.

The effectiveness of your promotion is the result of this subconscious effort. Depending on your general attitude, your knowledge of the subject at hand and the enthusiasm you display, your marketing effort will always have either a positive or a negative result. The latter is self-destructive and undoubtedly responsible for failures. It will always hold you back like a large lead ball chained to your ankle.

If you happen to be on the winning side of that equation, that's good. On the other hand, if you happen to be on the losing side and want to excel or if you wonder why you're constantly coming across stumbling blocks, then you should take a good inventory of yourself. Make a conscientious effort to improve in those problem areas that show up in the following two assessment tables.

At this crossroads, these two tables will guide you through an important challenge—the finding out how you stand in the electrical contracting marketplace and understanding the effect your personal traits, business skills, and technical skills have on your career and earning power.

Personal Assessment Table for the Productive Electrician

Code: R = Required/ beneficial to you P = Possess it now (8-10) M = Must Improve (1-7)	Self-rating. Use a 1 to 10 scale, 1 being low, 10 high		Productive Journeyman			Productive Serviceman			Productive Foreman			Productive General Foreman			Productive Project Manager		
	Before	After	R	P	M	R	P	M	R	P	M	R	P	M	R	P	M
PERSONAL TRAITS																	
1. Appearance			✓			✓			✓			✓			✓		
2. General Attitude			✓			✓			✓			✓			✓		
3. Punctuality			✓			✓			✓			✓			✓		
4. Enthusiasm			✓			✓			✓			✓			✓		
5. Assertiveness			✓			✓			✓			✓			✓		
6. Communication			✓			✓			✓			✓			✓		
7. Reliability			✓			✓			✓			✓			✓		
8. Foresight			✓			✓			✓			✓			✓		
9. Following Instructions			✓			✓			✓			✓			✓		
10. Straight-shooter			✓			✓			✓			✓			✓		
Personal Traits Total Score:																	
BUSINESS SKILLS																	
1. Time Management			✓			✓			✓			✓			✓		
2. Motivating Leadership			✓			✓			✓			✓			✓		
3. Decision-Making			✓			✓			✓			✓			✓		
4. Organizational Skills			✓			✓			✓			✓			✓		
5. Planning			✓			✓			✓			✓			✓		
6. Marketing			✓			✓			✓			✓			✓		
7. Negotiating			✓			✓			✓			✓			✓		
8. Purchasing			✓			✓			✓			✓			✓		
9. Accounting, Basic			✓			✓			✓			✓			✓		
10. Recordkeeping			✓			✓			✓			✓			✓		
Business Skills Total Score:																	
TECHNICAL SKILLS																	
1. NEC knowledge			✓			✓			✓			✓			✓		
2. Install and repair work			✓			✓			✓			✓			✓		
3. Plans and specs reading			✓			✓			✓			✓			✓		
4. Decipher schematics			✓			✓			✓			✓			✓		
5. Calculus, Basic Electrical			✓			✓			✓			✓			✓		
6. Layout Work			✓			✓			✓			✓			✓		
7. Motor Controls			✓			✓			✓			✓			✓		
8. Control Wiring			✓			✓			✓			✓			✓		
9. Knowledge of other trades			✓			✓			✓			✓			✓		
10. Safety regulations			✓			✓			✓			✓			✓		
Technical Skills Total Score:																	
Grand Total Score:																	

PERSONAL ASSESSMENT TABLE

Through the "Personal Assessment Table," you can rate yourself before and after you embark on your journey. Before you use this table, make a photocopy so you can use it later on for further evaluations. Or, instead of rating yourself, you may wish to consult the opinion of someone you trust. Use the "Before" and "After" columns for this purpose and date both events. As your score goes up, so will your value as a Productive Electrician.

You will notice that the right side of the first form is divided into five columns; each headed by the word 'Productive,' followed by a job classification such as Journeyman, Serviceman, Foreman, General Foreman, and Project Manager. Each classification column is subdivided into three smaller columns: R, P, and M.

Each imprinted check-mark in the R column indicates that the subject is either 'Required' or very beneficial for the productive fulfillment of that job classification.

For example, to qualify as a Productive Journeyman, as defined in this book, you must possess a score at or above the guidelines set for each skill or personal trait check-marked in the R column under the Productive Journeyman heading. This process is then repeated for the Serviceman, Foreman, General Foreman, and Project Manager.

The 'P' (Possess it now) column, can only be checked off if your self-rating for a specific subject is between 8 and 10. Any score less than 8 will force you to check-mark the 'M' (Must improve) column.

On this principle, as you improve in your weaker areas you will find yourself progressing toward becoming an accomplished Productive Electrician capable of holding any of the five jobs listed at will.

Your personal traits score should not be less than 90.

Your business skills score should not be less than 85.

Your technical skills score should not be less than 70.

To be at the top of the heap, your combined score should total no less than 260, at which level you can congratulate yourself on being an accomplished electrician who will always be in high demand.

Job Description and Qualification Table

Code: R = Required/ beneficial to you P = Possess it now M = Must Improve Industry Requirements	Subject Reference Chapter	Productive Journeyman			Productive Serviceman			Productive Foreman			Productive General Foreman			Productive Project Manager		
		R	P	M	R	P	M	R	P	M	R	P	M	R	P	M
1. Market yourself and the company	1,2,6,7	✓			✓			✓			✓			✓		
2. Deal with the customer (Owner/GC)	2,5				✓			✓			✓			✓		
3. Prepare for and attend meetings	2							✓			✓			✓		
4. Contract basics	3,8	✓			✓			✓			✓			✓		
5. Prepare/execute:Proposal/Contracts	3				✓									✓		
6. Service Calls	4				✓											
7. Work Orders	4	✓			✓			✓			✓			✓		
8. Change Orders	4,11,12	✓			✓			✓			✓			✓		
9. Contracts and negotiation	8,9				✓									✓		
10. Protect the job	10										✓			✓		
11. Maintain progress schedules	2							✓			✓			✓		
12. File legal notices	10,11,12													✓		
13. Job compliance	11				✓			✓			✓			✓		
14. Process shop drawings/substitutions	2,11,14													✓		
15. Maintain as-built drawings	2,11	✓			✓			✓			✓					
16. Submit contract requisitions	12													✓		
17. Mobilize the Job	13				✓									✓		
18. Lay out the work	13	✓			✓			✓			✓			✓		
19. Police limitation of contracted items	13,14	✓			✓			✓			✓			✓		
20. Redesign the work	13										✓			✓		
21. Labor: Crew make-up/replacement	13,14							✓			✓					
22. Material: Prepare bill of material	13,14	✓			✓			✓			✓					
23. Acquisition of lot purchases	13													✓		
24. Tools usage	13	✓			✓			✓			✓					
25. Cost control management	14													✓		
26. Weekly payroll reports	14	✓			✓			✓			✓			✓		
27. Compliance to codes/job inspections	2,16	✓			✓			✓			✓			✓		
28. Warehouse material	17	✓			✓			✓			✓			✓		
Total Classification Credits:		12			18			17			19			23		

JOB DESCRIPTION AND QUALIFICATION TABLE

This table gives you an overview of the five job classifications covered in this book. Each requirement attained is given a value of one credit. Through this chart you can determine what the industry expects from each classification, how well you qualify, and, all things being equal, what you're worth when compared to your local average pay-rate—should you ask for a pay raise or demote yourself? Or better yet, should you get to work on improving your present position?

The totals of each 'R' column represent the number of credits required to qualify for each job classification. Equate one of these totals to your local average pay for that classification and you will determine a credit pay-value. Multiply this value by the number of credits required in each of the remaining classifications and you'll find the average pay-scale for each classification.

For example, if a journeyman's average pay, in your area, is $16.50 per hour, you can determine that each credit-value is $1.375 ($16.50 divided by column 'R' 12 credits equals $1.375). Therefore, with each class being in direct relationship to the same level of expertise and the number of job assignments, we can further determine a basic pay scale for each of them. If that same journeyman's average pay is $12 per hour, then the credit-value will drop to $1.00

In the first case, however, the $1.375 credit-value should earn an hourly rate of $24.66 for the Serviceman's 18 credits; $23.29 for the Foreman's 17 credits; $26.03 for the General Foreman's 19 credits; and $31.51 for the Project Manager's 23 credits.

This equation is obviously based on full qualification of the minimum requirements shown in the table. For example, the journeyman whose rate is used for our equation must fully qualify on all 14 credits.

If the assessment table reveals that you are holding a position with less credits than those required, then you should multiply this lower number by the same credit-value to find your true hourly-rate worth. If you're paid more than you're worth, then your job is in jeopardy—not to mention the negative effect your lack of knowledge has on the company's bottom line.

Painful as it might be, be honest with yourself and do something about it because the security of your job or the future of your company may be in jeopardy. In essence, the message is clear: If you want job security and average or above-average pay, you must earn your position, not impose yourself on it.

To help you in this endeavor, the 'Subject Reference' column indicates the Chapters in this book where you can study a specific subject in depth. If need be, you may consider additional study that may help you with work habits and general attitude.

This table offers you the unique opportunity to evaluate yourself in depth and to target those areas needing improvement. When you have followed these guidelines and progressed to your ultimate goal, then be confident that your newly earned position is secured by the most powerful tool known to mankind—knowledge.

WHAT'S AHEAD

If we recognize that the successful growth of a company is ultimately dependent on the knowledge of its managerial team and its ability to focus harmoniously on the same objectives, then we can also recognize that the ultimate knowledge of the team members must be in synch with each other and with the company's ultimate objectives.

For example, the electrical contractor's resources, knowledge, and business skills will be far more effective if all the members of his managerial team worked by the same business principles he does; rather than by an array of methods based on bits and pieces of business skills acquired on a trial and error and most often experienced by one member only.

The most productive way to develop a harmonious team capable of focusing collectively and individually on the company's objectives is for all members of that team, including the owner, to adopt business methods of which each member is knowledgeable or has an opportunity to learn. Those methods must be all taken from the same source.

On this very principle, the following chapters have been adapted from *Electrical Contracting,* the author's operating manual for the electrical contractor and the basis for other works pertaining to the electrical contractor's managerial team, such as *The Essentials of Electrical Estimating and Selling the Job* and *The Office Administrator for the Electrical Contractor.*

The material ahead is relevant to the discovery of your relative position in the electrical contracting industry as well as to your general development as a Productive Electrician.

While some subjects may seem out of the realm of this work, you should pay particular attention to these extraordinary subjects, for they will help you understand the electrical contracting business as a whole.

Someday you may be called on to take over the electrical contractor's job and you won't want to miss that opportunity for lack of this basic knowledge. As stated in the beginning, this book offers you the unique opportunity to be the best you can be in your industry.

What's ahead in this manual is a combination of educational subjects relevant to the previous two tables and to your general development as a Productive Electrician. Some chapters will delve into the job of the electrical contractor in order to acquaint you with some of the problems he has to deal with to run a successful operation.

You should pay particular attention to these chapters for they will help you understand the electrical contracting business as a whole.

Much of the following material is adapted from the author's *Electrical Contracting* book.

It's the author's intention to create uniform methods of teaching whereby the Electrical Contractor and his managerial team all work in harmony and are able to focus on any given subject in a synchronous mode. The Productive Electrician and his employer should be able to communicate with each other on the same level. The employee and the employer will know what to expect from each other or how to respond to a specific event on the fly without much ado.

Chapter 2

MARKET YOURSELF AND THE COMPANY YOU WORK FOR

THIS AND OTHER CHAPTERS

This chapter deals with the marketing of yourself and the company you work for to the customer with whom you come in contact with, whether directly or indirectly. The first requisite of a productive electrician is to always promote himself and his company to the industry he serves. This kind of loyalty is an unwritten rule that applies to all employees.

While several references in this and other chapters that follow can easily be applied to the way you market yourself to your employer, their basic purpose is to teach fundamental business skills that will help you become a valuable member of your company's managerial team.

If you need to review the marketing of yourself to the electrical contracting industry, then go back to the Introduction and Chapter 1.

MARKET YOURSELF

Historically, illegal and unqualified contractors have often enticed prospective customers with the lure of low prices, setting a trend that greatly affects the legitimate electrical contractor. Shady contractors with slick presentations quite often take away jobs from qualified contractors who have yet to learn the art of effective marketing.

As a serious electrical contractor's representative, you must recognize and deal with this trend, as well as with other competitive elements. You must follow or help generate a comprehensive marketing plan that allows you to market yourself and the company; not only to prospective customers, but to the industry as a whole.

Those who underestimate the value of a marketing strategy usually leave behind a trail of misleading impressions that foster endless negative repercussions, affecting the company they work for and all they do.

FIRST IMPRESSIONS

When you promote yourself, either by personal contact or by printed matter, it is important to leave behind a favorable impression. The visual impact you create, followed by your spoken words, are judged and forever recorded in people's minds. That impression is also labeled with their perception of who you led them to believe you are and what you led them to believe you do. This label is usually a product of your presentation, and, once set in their minds, it rarely changes. First impressions are inerasable. They will remain unchanged. From that moment on, every subsequent contact you make with them will trigger their first impression of you.

If a person first presents himself as a serviceperson to a customer and later on wants to change his image to that of a businessperson, for example, he can never upgrade his rank in his customer's mind-eye unless he first presents himself as a well-rounded businessperson even when performing service work. The image he creates first is how that customer will label him no matter how hard he tries to change later on.

Regardless of the rank you choose—serviceperson, journeyman, foreman, general foreman, project manager, or whatever else you may choose—you should never make careless or un-businesslike presentations. Unprofessional presentations create long-lasting damage to your image, and lead prospective customers to either look elsewhere for a better contractor, or to treat your company like that of any other shady contractor unworthy of their business, unless their risk is well compensated. When the latter takes place, suddenly your lowest price is not low enough, no matter how hard you try to make the sale.

But favorable impressions serve only as door openers to better opportunities. They are not signed contracts. They are to be treated no differently than telephone calls generated by solicitations. What happens with these opportunities is entirely up to you and your ability to sell and close contracts. Once you take a job, the quality of your company's performance will then complement your other virtues. This determines the type of relationship you establish between your company and the customers. Will they call you back?

FIRST MEETINGS

In making sales presentations, be aware that the first question in prospective customers' minds is: "Who are you? Who's your company?" And unless you're ready to answer promptly and satisfy this and other pertinent questions, the presentation can easily turn into a very uncomfortable and non-productive meeting.

In a well-planned marketing approach, your promotional material has to lead up to your personal presentation. This approach will familiarize the prospective customer with you and your company and pave the way for productive sales calls. When you finally meet your prospective customer, there will be no question as to who you are or who your company is. The job will then be the meeting's focal point.

However, when the occasion arises that you have to make "cold calls" in person, one sure way to avoid embarrassment and win the prospective customer's confidence is to be ready to answer all questions and to overcome the "Who are you?" question within the first few seconds of the meeting. This is done, not by magic, but by an efficient presentation adhering to some basic guidelines. For a productive electrician, the things-to-do list is simple and concise:

Personal Appearance

Most people observe people from their feet up. Good clean shoes, neatly groomed hair and complementary attire will do the job.

Show yourself to be ready to do business by carrying a briefcase or a well-organized notebook.

GENERAL ATTITUDE

- Be positive and enthusiastic about the topic at hand. Show your expertise and genuine interest in the project.
- Pay close attention to what your prospective customer is saying.
- Express your opinions freely.
- Do not use his telephone for any other purposes than your presentation. When you have to, be very discreet.
- Converse with your prospective customer.

- Humor, when out of place, can be costly. Jokes and other trivial conversation should be left for the water-cooler breaks.
- Do not hide your traits; they are your assets and you should appropriately promote them.
- Do not throw ice cubes on a warming-up relationship. In other words, do not use phrases such as "Yes, Sir" or "Yes, Ma'am" or the like. There is nothing wrong with addressing customers by their names. When someone is warming up to you, it is because he begins to trust you and wants to get closer. A too formal conversation can distance your prospective customer to an irretrievable corner. Seize the moment, and share his enthusiasm in the most cordial and friendly way. If he warms up to you, respond in the same way. Your prospective customer, before he signs the contract, wants and needs to feel comfortable with you and your company.
- Do not make constant reference to the National Electrical Code. Use your mechanic's expert opinion. Remember the N.E.C. requirements are minimum standards of good workmanship. Most projects are designed and built above minimum standards.
- Do not criticize the job design's criteria until you have learned the reasons why it was designed that way. Then if you feel you have to criticize, be very tactful.
- No snow jobs: You are better off saying nothing than to fake a compliment.
- Also remember that telling someone you or your company is licensed adds little value to your presentation. You are expected to be licensed. On the other hand, to tell someone you do work above industry standards adds considerable value to your presentation.

MARKET YOUR COMPANY

Before we start on the subject of marketing your company, we need to address a misconception that concerns many industries. The primary intent behind marketing a company (not to be confused with marketing a company's product or service) is to enhance and present the company as a whole to prospective customers.

Misleading promotional campaigns are the main cause for many consumer losses with costly backlashes to the promoting company itself. For example, a growing company that is working hard at creating demand for its services makes unjustified changes to their marketing plans in midstream. Many executives consider natural growth to be too slow. When that happens, they begin to doubt sound marketing approaches, they disregard their company's ability to produce and, at the flip of a coin, embark on a new and more aggressive campaign, thus generating more sales than the company can handle. This misleading campaign adversely affects the customer and the company as well, especially when orders are mishandled.

We need not be seasoned marketers to spot the problems facing these companies. They simply are not set up to handle overnight growth; for that matter, very few companies are. The additional workload can easily put them out of business.

Marketing plans have to be laid out with care. Their implementation must be executed with greater care. And their results need to be patiently nurtured for long-term growth.

Promoting a company for a job for which it is undercapitalized or not equipped to handle, will burden the company's natural growth. Natural business growth is achieved by letting the business grow at its own pace and on its own merit. This means the company should undertake only profitable jobs that it can handle with ease. Consistent profit-making gives confidence in a marketing plan, which in turn nurtures good organization—an essential element in the natural growth of long-term plans.

Forced business growth, on the other hand, besides being a gamble, will lead to disorganized plans, with daily tasks run on emergency bases, and new personnel training reduced to almost nothing. When a company puts itself in that position, it will soon find itself having neither the resources nor the organization to complete the work.

To prevent such a situation, once a company has reached its workload capacity, no matter how hard it is to turn down a customer, or what seems to be a profitable job, the company has to turn it down. The company can't take on additional work just on gut feeling. Decisions must be based on facts and figures (See Chapter 19, "Controlling The Job.") and a good plan to expand its work capability. Some may think this rule applies only to manufacturers or larger contractors. This could not be further from the truth. It applies more to the beginner than it does to large established companies.

As wise and successful businesspersons, the principals of the company must update themselves on its financial status and production capabilities on a daily basis. This knowledge will give them the confidence to make wise decisions. With it, they may take calculated risks and undertake additional work, or they can build the courage to turn down jobs that, in their opinion, regardless of anticipated profits, are a sure liability to the company.

If you worry about the company losing customers by turning down work, don't worry. In such cases, the first thing the company has to consider is its responsibility toward its employees and itself. In these situations, the company may turn its predicament into a powerful marketing tool. Explaining to your customers why you are turning down their work will only strengthen your business ties. Everyone wants a winner. You will be planting a very good seed for the future, thus creating greater demand for your company. This customer will not only call on your company early for his next project, but will influence others to call on you as well by talking about your integrity and the successful operation of your company.

Good, effective marketing plans grow at their own pace. After all, the seeds you plant for your company are the seeds for your personal future as well. Because so much depends on them, when we sense a delay or decline in sales, it is a common syndrome among planners and executives to start doubting the very plan they created and implemented. These are moments in life, however, when a company has to believe in its managerial team and patiently stay put.

This is not to imply that marketing plans are written in stone. If a company feels it has given its marketing plan a reasonable chance to work, by all means it should make adjustments as it goes along. But it should be done prudently, because neither you nor anyone else in the company can predict where, when or how many seeds are about to germinate. A company should make changes, but never stray from good marketing fundamentals.

A COMPANY MARKETING PLAN

For an electrical contractor, implementing an effective, comprehensive marketing plan usually does not add to his advertising costs. At times it even cuts his budget. Chances are, he is paying for some form of advertising—for example, his company stationery, truck lettering, specialty items, yellow pages and the like. Most often, the only thing he needs to do is rework and synchronize them with those missing elements that are needed to achieve his company's objectives.

A common alternative to a comprehensive marketing plan is a random plan. As the title implies, it's a combination of traditional marketing elements, most often copied from others, and the adoption of randomly selected, ineffective bits and pieces of advertising and specialty items.

At times, such plans may give the appearance of a comprehensive marketing plan. Ironically, though, with a strong compulsion to invest in advertising on one hand, and cut expenses on the other, a contractor creates costly and ineffective marketing plans, thus squandering good promotional dollars.

To convert ineffective advertising items into effective elements for a new comprehensive marketing plan, a contractor should begin, for example, with his truck lettering. Next time he's lettering a truck, he should make sure the overall layout sends the message he wants. Next, he should do the same with his stationery and so on. (How to do this is detailed later in this chapter.)

In comparing cost versus effectiveness of individual marketing elements, a contractor will find, amazingly, that the least expensive are the most effective. A well-written company reference letter, for example, or the implementation of standard procedures aimed at the enhancement of the company's overall performance, are the most common among these items.

Starting out in business, most contractors pay little or no attention to image setting and visual impact. In many instances it's the last thing on most people's minds. However, for the someday-to-be-successful-contractor, starting up is the best time to follow his visions and plan where he wants his company to be five, ten or twenty years down the road.

The implementation of a marketing plan, for those who are committed to go the full mile, is a serious matter and as such should not be left in the hands of others. Copywriters and commercial artists can only create from what a contractor shows and tells them. If a contractor wants his experience and wisdom reflected in their work, then, contrary to what he thought or to what he's accustomed to, he must provide them with specific guidelines as well as his visions.

Following is a list of the major marketing elements pertaining to the electrical contracting business. The commentary notes are designed to acquaint the contractor with acceptable guidelines for creating effective marketing campaigns. A contractor should be directly involved in the future of his company. He should study these notes in detail, for they will give him, and you, the basic knowledge to define effective marketing plans. Learning these elements will help you in your endeavor to become a productive electrician.

Logo

A logo is the catalyst of the company. It is the image setter that tells what the company is, what it does, and it must portray the same message regardless of its size and color. While it should be unique, it does not have to be clever. The best design for a logo is simple in color and layout. A logo, due to its different applications will sometimes be reduced to a very small size, or enlarged greatly. It will also be photocopied in black and white, especially in today's world of facsimile (fax) machines. If it's loaded with many or tiny details, and too many shades of color, those details will be lost when the logo is reduced or be out of proportion when it's enlarged. The most effective are graphic logos.

Truck Lettering

The best and most inexpensive mobile advertising space is the free space on a truck's outside panels. A contractor must use all sides of that truck, including the top when necessary. But as often as their trucks are seen on the road, many contractors waste this valuable advertising space. For example, a majority of them rarely use the rear or the front panel and never the top.

Using the top of a service truck is very effective advertising, especially when serving business and residential communities with two-story buildings and larger. When a truck is parked between other vehicles, the top panel has a greater audience than the side panels. The people on the ground level are usually in a hurry, while the people up above; attracted by the novelty, have the time to write down the company's name and phone number. If a contractor decides to use the top panel of his trucks and must also place the ladders up there, then share the space between the ladders and the sign. He must move and stack the ladders to the side opposite the sign.

Lettering the rear of the truck, because of the long-time exposure it has for those who follow your truck, can't be neglected. Too many companies, to save a few dollars when ordering lettering, are wasting this valuable space.

The general guidelines for truck lettering are as important as the location of the lettering on your truck.

Do not clutter the lettering with too much text. Every word used, in its proper place and size, has to say something of value (No ego trips here).

The size of each line should be consistent and proper. The name of the company should be 100%, the telephone number 75%, a slogan 30%, and the rest of the text, if any, should be proportionally smaller.

License Numbers and The Like

License numbers, though they are required and though they reassure the public of the company legitimacy, they don't carry much advertising weight; no matter how large the letters are. They should be treated as normal text, 30% or less of the company's name.

Telephone Numbers

Whenever possible, a contractor should void the use of multiple telephone numbers—for example, one for each location he works in. This type of listing tends to clutter the truck lettering, and takes the reader's attention from the message. If the intent is to make it convenient for anyone to call the company from remote locations, then adopt an 800 or 888 number. The toll-free call also serves to overcome people's reluctance to call out-of-town contractors. It's a good business rule to pay for business calls generated by the company advertising. In business, a contractor will always pay for leads one way or another.

Color and Font Typefaces

The use of too many colors should not be employed; the maximum should be three colors. The same applies to font and typeface styles for the letters.

Company Logo

A well-displayed company logo on a truck or sign will pull more than any logo printed on business cards and letterheads put together. This is because the truck or sign is exposed to larger audiences for greater periods of time than a business card or letterhead, which are usually seen only by those that are personally solicited or the company deals with. This is only a comparison between outdoor signs and general stationery, and not a recommendation to remove the logo from the company stationery. The logo is important in that it delivers the impact for all the company's visual marketing tools. The comparison here is made to encourage the placement of the logo on trucks and signs, a neglected item by many contractors.

Therefore, next time a contractor wants to cut the truck lettering costs, he should shop for better prices rather than omit the logo. Besides, there are most likely many other items from the truck lettering layout that are ineffective that a contractor can delete, not only bringing his cost down, but also improving the overall impact of the company's message.

Paint or Vinyl Lettering

For all his sign needs, whenever possible, a contractor should use cut-to-specifications vinyl letters instead of hand-painted lettering. The vinyl lettering lasts longer, better resists the effects of weather, and does not discolor as fast as painted letters do. The overall look is also superior.

If budget is his problem, many vinyl lettering suppliers will cut his letters and logo to his specifications. With a little guidance from them, he can install it himself and save enough on labor to afford lettering for all five sides of the truck plus the company's logo.

Job Site Signs

The same rules that apply to truck lettering basically apply to any form of sign lettering with few exceptions:

- Unlike truck signs, job site signs are fixed, and passing traffic must get their message at a glance, thus limiting the text a contractor can place on them. The logo and company's name should be the most prominent elements. Job site signs, when well proportioned in line size, font, and color, pull a good share of advertising load. Because they are placed near a job a company is doing, they act as testimonial to its success and are therefore very effective. Placing one sign at each job site around town will give the company the best credentials it can have.

- A contractor should avoid popular layout signs where *"Another Job By"* or *"Electrical Work By"* lines are larger than the company's name. If the sign is next to a construction site, it's obvious that the company is doing that job. To add any other line will obscure the company name and logo.

- A contractor should keep the same lettering proportions for his sign lettering that he used on the truck: name and logo 100%, telephone number 75%, all other lines 30% or smaller. He should check that every word says something of value. He should not clutter the sign with unnecessary words or lines and he should make use of good contrasting colors.

Uniforms

Uniforms are a necessity for service-oriented companies. Their color scheme and lettering should stay in harmony with the rest of the company's attire. A contractor should use identification tags for each serviceperson instead of imprinting their names on the uniform.

Identification tags

Identification tags add prestige to the company. The customer feels the company has taken the time, which a contractor should, to individually check and identify its employees, especially when the employee photo is included. Identification tags make it easy for exchanging uniforms and are a requirement in most security-sensitive areas. In a marketing reference letter, the mention of the company using uniforms and identification tags for its servicemen many times is the difference between losing and getting a job or a maintenance contract.

Hard Hats

Hard hats are good safety devices. OSHA requires them on all job sites. From a marketing point of view, they are very effective. The next time you see a cover page featuring a contractor at work, note the two main elements of visual impact. The hard hat and the company's name and logo both enhance the publication's feature story, as well as the contractor. Visualize the same photos without hard hats and you will understand the power of visual impact.

Indisputably, hard hats make a statement about the company. Those who think it's inconvenient or unimportant to make their employees wear hard hats should consider safety as well as marketing advantages. The use of hard hats makes good business sense since, for relatively little cost, it creates a safer work environment with fewer accidents and lower insurance premiums while giving the company greater visibility.

COMMUNICATION

Stationery

A letter is the physical messenger of your thoughts to whomever you want to address them. Its appearance dictates the impression you will create on the reader.

Many volumes have been written on business letter-writing content, text formats, and layout including crafted letters for any business occasion. If you need help in business letter writing, you are encouraged to follow the guidelines as laid out in specially crafted business letters. They are an excellent learning tool.

There are certain conventions used in letter writing that are well established, yet they are flexible enough to allow us to communicate exactly what we want to our reader. If we consider the appearance of a company letter, the stationery, format, length, and envelope the reader will be drawn to it. Once the reader gives a letter his or her attention, your message is sure to get through.

The Envelope

The appearance of the envelope adds to the overall professional appearance of the company letter. With the exception of using the two-letter abbreviation for the state, the address on the envelope should appear exactly the same as the inside address of the letter.

The Letterhead

The letterhead usually consists of the company logo, the company full business name and address, license number, phone and facsimile (fax) number. The information included should be uncluttered and readable. The design should be simple enough for the reader to find the information he or she needs without being distracted from reading the rest of the letter.

The facsimile number is an additional (electronic) mailing address and as such, it belongs with the address, possibly on the same line and right after the zip code. Contrary to conventional practices, it does not belong next to the telephone number. The telephone number clearly belongs by itself, clear of clutter.

Business stationery usually is white or some other conservative color. The standard size of the stationery is 8.5 by 11 inches.

The letterhead is always used only as the first sheet of a letter. If the typed letter is more than one page, a plain sheet of paper matching the letterhead should be used for subsequent pages.

The length of any letter affects its appearance. Customers who receive a lot of correspondence every day are not going to react favorably to lengthy three-page letters that could have been written in one page.

Come right to the point in your letters. They should be concise and limited to one page if possible.

REFERENCE LETTER

A reference letter is the company passport to better accounts and better jobs. A reference letter consists of the company data sheet (see below) with a reference sheet listing completed and active jobs with their respective customers' names and phone numbers. A contractor, if he wishes, may list the project dollar value. Of the two sheets, the data sheet will be the more important to a prospective customer.

Company Data Sheet

A company data sheet should be written on its letterhead, and it should start with a short welcome paragraph and include the following items:

- The contractor's electrical license number, the issuing municipality or agency, and the territory the contractor is licensed for.
- The federal identification number known as FEIN (Federal Employer Identification Number).
- The number of whatever municipal, county or state license(s) is required to operate in that territory.
- The Workman's Compensation and Employer Liability's carrier, including policy number.
- The commercial General Liability's carrier including coverage and policy number.
- The insurance agent, including contact name and phone number.
- Any association with electrical contracting councils or boards and the like, including years of membership.
- Operation. A short statement of company operating territories, including remote shops or branches, when applicable.
- Management. A short statement of the company management philosophy and why its method of doing business will be beneficial to the prospective customer.
- Banking. The bank the company deals with, including contact person and telephone number.
- References. A contractor should use this final statement to refer the reader to the next page, "The Company's Reference Sheet." He may write: "The following is a partial list of customers . . ." or any similar statement that will make the reader aware that there is more to his presentation.

Company Reference Sheet

- In this sheet a contractor may list his completed projects by category: Residential, Commercial, and Industrial by territory or both. The combinations are endless; however, he should not overdo it. He should stay within one page. Normally, if he presents his company well in its data sheet, the reference sheet becomes secondary and he need not be elaborate.

- A contractor should insert a P. S. note at the end of the reference sheet stating something similar to this: "For further information please call Mr. John Doe at (123) 456-7890."

John Doe Electric Co.
Licensed Electrical Contractor No. E-12345

123 Main St, Our City, USA, 12345, Fax (123) 456-7890

Call us . . . (123) 555-6400

—Company Data Sheet—

Welcome to John Doe Electric Co,

The information contained in this letter is to acquaint you better with our company. We stand ready to assist you with your upcoming project. Just call us.

License and FEIN numbers:
Electrical license number E-12345 held by Mr. John Doe.
Federal identification number 65-0000000

Insurance:
Workers compensation and employers liability; XYZ Company, policy #407-3333-01
Agent: Independent Associates; Mr. Smith (123) 234-45678

Associations:
Electrical Council of USA
North-star Association of Electrical Contractors
John Doe Electric is a registered XYZ power and light company for Kings Counties

Operation:
The Company is licensed to operate throughout the state; however, it maintains its activities within Kings counties.

Management:
The Company manages its work through four independent branches. This method provides our customers with competitive prices, person-to-person relationships, and better job control. The branch manager is directly responsible to you for the execution of the job.

Banking:
Any-Bank 3333 Main Street, My City, USA; account manager Ms. Doris (123) 777-3434

Jobs References:
For references see the attached "Jobs Reference Sheet."

John Doe Electric Co. - Reference Sheet

RESIDENTIAL

Chaney Residence	Bani Construction	(123) 555-6767
17031 Ocean Blvd.	2569 South Stewart	
Any City	Any City	
Lazo Residence	P.L. Douglas Construction	(123) 555-8856
2173 S.W. 13th Ct.	611 Medallion	
Any City	Any City	
Ladder Homes Inc.	B.C. Builders	(123) 555-9966
5205 West Hill Rd.	7788 Paris Rd.	
Any City	Any City	

COMMERCIAL

Tuscany Grill	Jankin Contractors	(123) 555-2233
1203 Main Street	40111 Atlantic Ave.	
Any City	Any City	
Baron Sports	XYZ Development Inc.	(123) 555-4510
34 Washington Street	5569 Monroe Street	
Any Town	Any City	
Merry-Go-Round	Store Builders Inc.	(123) 555-7790
2500 North Broadway	7774 Geneva Blvd.	
Any Town	Any City	

INDUSTRIAL

Morris Paint Co.	W.P. Ford Corp.	(123) 555-9999
Route 22	8934 Western Lane	
Any City	Any City	
Falcon Foundry Corp.	By Owner	(123) 555-0087
5677 Babylon Road	5677 Babylon Road	
Any City	Any City	

For further information please call John Doe at (123) 555-6400

PRINTED AD

The general guidelines for effective visual impact a contractor has learned here also applies to printed ads in publications, telephone directories, circulars, brochures and specialty items. These guidelines, combined with his marketing objectives, will help him create the most effective layout for his next printed ad.

PERFORMANCE

As a promotional tool, nothing has a more lasting marketing effect than good company performance. The quality of its performance perhaps will not generate a great deal more business at first, but a contractor can be certain that in time it will become his company's hallmark.

To gain such a reputation takes conscientious and collective effort by every member of the company. A contractor should adopt policies that will deliver quality service with integrity, for he will be on his way to an unstoppable success. The following topics pertain to this subject:

Punctuality

The worst feeling you can have is to walk into a meeting and instead of saying *"Good morning,"* or *"Good afternoon,"* you have to say *"I am sorry I'm late,"* or make up excuses for being late. The worst part is not your feelings being hurt, but your reputation as a responsible individual and that of the company you work for being seriously damaged. Frequent incidents like these adversely influence decision-makers.

If you are a habitually tardy person and wish to experience what others experience when you are late, just think back to a situation when someone was late for an appointment with you. As the time approached, you began to check your wristwatch. At about five minutes before the hour, you began to formulate doubtful thoughts about that person. At the top of the hour your doubtful thoughts changed to unprintable language that got progressively worse with every passing minute. Fifteen minutes past the hour you had lost all respect and begun to consider him as one you take lightly.

Now that you know how others feel and think about you when you are late for an appointment, think of the effect this has on your image as a responsible individual. No amount of *"I'm sorry I'm late,"* will ever recover the damage. Show up on time and be considerate of other people's time.

Say it straight

When you say something, mean it and say it straight. Don't underestimate and double-talk a customer; most often it backfires on you. Be a straight shooter.

Don't wait for them to call you. Get a head start.

Once a contractor has been awarded a job, he's expected to take full charge. He has to keep an eye on the job's progress. He must not wait for the customer to call him. The idle time, from the award until the job goes into full swing is the best time for a contractor to set up and get a head start, especially in alteration work. A contractor should mobilize the job and physically do whatever work he can at his own pace. Remember the customer is not the electrical contractor; the contractor is. The customer will usually call him when he thinks the job is ready for him. At that point, usually he will wish he had been there sooner. Staying attentive to the job's needs prevents costly mistakes. It also builds a good working relationship with the customer.

Get timely job inspections

In our industry, passing scheduled inspections is crucial. When a contractor neglects to call for an inspection, or fails one, the entire construction schedule can go off course. If he does this a couple of times, he will receive threatening letters from his customer asking him to shape up or ship out.

Before calling official inspections, a contractor should be certain his work is ready and he or someone from his company meets the inspector at the job site. Inspections go more smoothly when he is available at the inspection site to answer pertinent questions.

FOLLOW UP ON PAPERWORK

Timely paperwork processing is as important to contract administration as passing inspections is to construction schedules. Procrastination in this area is common among failing contractors. A contractor's efficiency directly affects job progress and collections.

Compliance with contracts, depending on the owner's requirements, can generate a great deal of paperwork. However, regardless of how much paperwork there is, it is as much a part of the contractor's contractual obligation as is the installation of conduit and wire. The only difference is that a contractor may easily get away from installing a certain conduit run, but he will not get away from processing any paperwork, especially if the customer needs it to satisfy the owner or a bank.

In our business there is no room for procrastination in either aspect of the work. Installation and contract administration are essentials to successful contracting. If the contractor does them efficiently he will have a well balanced, performing company.

The following items, when they are part of the contract, require special attention:

Submittal

Any documentation requiring the contractor to evidence that: (1) material and equipment supplied by him are in compliance with plans and specifications and (2) the contractor complies with the general provisions of the contract. Customarily, these submissions are referred to as "submittals."

Submittals are usually transmitted for approval to the architect through the customer (the general contractor or the owner). Submittals are to be transmitted within a certain number of days from the date of "Notice of Commencement." The architect, in turn, has a certain number of days to either approve or reject a submission. This simple procedure, due to different contract requirements, most often gets complex and time consuming. For example, the circle of submissions and approvals, for any one item, can expand to include the manufacturer, the manufacturer's representative, the supplier, the contractor, the general contractor, the owner, the architect, the consulting engineers and other agencies having jurisdiction over the project. Therefore it behooves a contractor: (1) to promptly transmit his submittals for approval and (2) not to release any work or material that is subject to those submittals until he is in possession of their approval.

Staying within deadlines, despite this circular correspondence among the contractor, the general contractor, and the architect, is a matter of survival. If a contractor wants to comply with the terms of his agreement and survive as a successful contractor, he cannot procrastinate. Its ramifications will become more evident in later chapters.

At this juncture, however, it is worth noting that punctuality can directly affect profit. A delayed approval of certain material can well disrupt its shipment and subsequently the job productivity.

The following is a list of the most common submittals:

- Progress Schedules

- Progress Reports
- List of Material Suppliers
- Shop Drawings
- Change Orders
- Material Samples
- Monthly Requisitions and Applicable Releases of Lien
- Payroll Reports
- Safety Report

Material Release

When the shop drawings are approved, a contractor should promptly forward them to his supplier along with his release order. Procrastination in placing an order or in releasing approved material and equipment delays the job just as delaying any other paperwork would.

A contractor should bear in mind that many manufacturers and their agents, during vacation, shut down their entire operation for extensive periods of time. Months before their scheduled vacation they begin to restrict shipping commitments. Their vacation schedules rarely are taken into consideration in any job progress schedule, thus leaving the contractor with the responsibility to coordinate purchases and deliveries.

If a contractor acts in a timely manner, within the terms of his contract, and meets all deadlines, there can never be any liability assessed to him for situations beyond his control. If he fails, however, there can be enough back charges to dissipate his profit and more.

As-built Drawings

As-built drawings are to be updated in the field as the work is being carried out. When a contractor neglects this simple task, as many do, at the conclusion of the project it becomes the most common barrier between him and his final payment.

JOB SITE CLEAN-UP

Job site clean-up, unless otherwise stipulated in your contract, is part of the job. Failure to comply with it more often than not results in large and unexpected back charges. If it's part of the contract, in order to prevent unexpected back charges, a contractor either agrees to a fixed amount he will be charged for the service before the job starts, or he does it himself on a daily basis, as it may be cheaper. Under no condition should he neglect this item of work; it creates many worthless arguments at the tail end of a job.

MARKETING RESULTS

If we want to guess we can safely state that nine out of ten contractors who would read this chapter, for one reason or another, will take exception to most of its marketing approaches, especially about clean-up and punctuality. Their minds run off with such thoughts as, *"I always get away with clean-up and stuff like that. . . why bother?"* These individuals undoubtedly will have a similar answer for every exception they take in life.

The fact, however, is that large numbers of contractors fail and go out of business. And if we speculate one more time with statistics, we can say these contractors are the same nine-out-of-ten who think they can get away with that *"stuff,"* and simply refuse to subscribe to good business practices. The remaining one out of ten, those that subscribe to good business practices and sound marketing fundamentals, are bound to succeed.

In conclusion, if a contractor truly wants success, he must take responsibility and deliver complete jobs, and that includes all the related paperwork. Any other practice is counter-productive. If he feels he can't handle jobs with extensive paperwork, he should decide before he signs the contract, or, for that matter, before he bids on them. A good indication as to the extent of paperwork is the type of owner. If he wants to be certain, he should check the entire bidding document.

Live up to contractual obligations. What a contractor will gain in reputation by far outweighs the few dollars he may save in cheating on trivial *"stuff."* He has to think of this cost as a good marketing investment.

In marketing, the main objective is to create demand. In our industry demand is created by accumulating as many favorable impressions as possible with those who are in a position to influence others and those who will directly decide on the next job award. This is one of the secrets of successful contracting.

When a contract is awarded, a contractor should seize the moment. With it, and with every other contract he undertakes, he has a unique opportunity to create greater demands for himself and for his company. A contract performed on time is the ultimate and most inexpensive marketing tool by which a contractor can establish himself as a competent and trustworthy contractor.

Chapter 3

CONTRACT BASICS

PRINCIPLES OF CONTRACT

Before you start selling the company services, we need to review a few basic principles that affect the closing of successful contracts. How we define a successful contract and how it comes about are of great importance to the development of a sound marketing approach.

To begin, a successful contract is one that a contractor carries out profitably and gets paid for in full, and the satisfied customer recommends the company to others. The main components of this contract are: 1) the customer solicitation for work he needs done, called the "Invitation to Bid," 2) the contractor's response to that solicitation, called "offer" or "bid" and 3) the customer's acceptance of the offer. When these three elements are fairly set and communicated with clarity, a good contract will arise; if not, the contract will lead to a troublesome business relationship, the very thing a contract must avoid.

The following definitions will help you explore these elements further:

The Solicitation

The solicitation, or bid document, details the scope of work the customer wants done. When a contractor deals with a lay person, it's to his advantage to exercise his technical and business expertise to guide the customer toward his goal. When he quotes jobs, with bidding documents prepared by experts, such as attorneys, architects and engineers, his focal point should be the documents' fine print, and not the job details. He may discover that parts of the proposed contract don't agree with his policies, or he may not be able to comply with some special requirements. When that occurs, unless he can rework these discrepancies with his prospective customer at the start, he's better off not bidding on the work. It takes courage to walk away from potential work, but it is only stupidity to knowingly walk into a lion's den.

The Offer

The offer, or quotation, is where a contractor defines his terms: In his proposal he must express with utmost clarity what it is that he wants, how much, and what he will or will not do.

The Acceptance

The acceptance, while it's referred to as the owner's acceptance, is reciprocal when the contractor and the customer sign a contract.

With these three elements defined, a contractor can begin to formulate the basic tools he'll need to sell successful contracts. Work Order, Service Call, Quotation, Proposal, or Proposal Contracts and the like, when well prepared in accordance with the principles described here, are documents that will protect the contractor, as well as the customer.

If a contractor wants to conduct smart business, your printed forms must say plainly what is correct for your type of business. Business smarts dictate avoiding troublesome contracts. For clarity's sake a contractor must use written words and business forms that people can easily understand, and, when necessary, readily refresh their memory. The right paperwork is the contractor's best protection. You'll be surprised how temptation is averted and how well each party performs when they know they are bound by a fair and valid contract.

PROPOSAL / CONTRACT FORM

The Proposal/Contract form shown in this chapter employs the very principles explained earlier. It contains the basic elements needed to enter into successful contracts. This form is introduced here first because, as is, it's usable in both activities: Service and Contracting Work.

The comments that follow on how and why blanks must be filled in are as crucial as the format itself. All forms shown in this manual have been and are successfully used by established contractors. Through years of experiments and through numerous revisions, their format has been reduced to a bare minimum. Each clause requires information that, when obtained from a prospective customer, will give the contractor necessary and efficient documentation to help him carry out contracting tasks with ease. It's therefore on this principle that a contractor should invest the few extra minutes that it takes to fill out these forms entirely. When he does, he'll make a considerable contribution to the success of his business.

2

1

John Doe Electric Co.
Licensed Electrical Contractor E-12345

123 Main St, City, USA, 12345, Fax (123) 456-7890

Call us . . . (123) 555-6400

Proposal /Contract
No: _____

Customer	Job
Contact:	Contact:
Name:	Name:
Add:	Add:
City/St/Zip:	City:
Phones: Hm: Wk:	Owner:
Beeper: Fax:	Job Phone:

4

John Doe Electric, referred to as "Contractor," proposes to furnish labor and material as specified herein for $_____ payable as per contract breakdown as submitted by the Contractor prior to commencing work, or as follows:

3

____ % = $ _____ down _____

____ % = $ _____ upon _____

____ % = $ _____ upon _____

6

Contractor's Rep: _____ Signature: _____ Date: _____

5

If this proposal is not accepted within 10 calendar days, then, without notice, the Contractor may withdraw it anytime thereafter.

7

Scope of Work

Install electrical wiring as per plans and specifications, if any, dated: _____ and prepared by _____ Identified as Drawings No:_____ revised on _____ with the following exceptions:

1

2   **(This and other forms may be downloaded online at www.theestimatingroom.com)**

3

8

Customer Acceptance

The price, specifications, and conditions set herein are satisfactory and hereby accepted. I authorize the contractor to proceed with the work as specified herein. I shall make payment as outlined above. Unpaid balances are subject to 1. 5% per month interest until paid in full. Customer agrees to pay all of the contractor's costs related to the collection of any sum due, including legal fees and other applicable expenses.

Patching, painting, site restoration work generated by this work shall be done by others. Lamps, appliance cords, and caps shall be supplied by others. Power company charges, municipal permit fees, and contractor's processing fees, if any, are not included in this proposal. The electrical permit shall cover only the work contracted herein. The cost for removing any existing violation is not included in this proposal. The Contractor, when required, for an additional charge to the customer, shall apply for an electrical permit to cover said extra work. Changes to the scope of work shall be written as "Change Orders" and executed by both parties prior to commencing any work on said changes.

Customer please note: When this Proposal/Contract exceeds $5000, in addition to the Contractor Rep's signature it requires the Contractor's owner/president signature here: _____, Pres ___/___/ 20___ or the contract shall be considered null and void.

I _____ am authorized to accept and sign this contract because I am the customer named above, or I am acting for the customer as his agent.

Customer Signature: _____ Title:_____ Date: _____

Figure 1: Proposal /Contract

Notes to Proposal/Contract

1 **Proposal/Contract Number-** A contract number is essential for cross reference with future documents such as Change Orders, Submittals, Payment Requisitions and other legal documents. A contractor should be consistent and use whatever numerical scheme best fits his operation.

2 **Customer-** In all forms it's essential to identify the person responsible for paying the bills as the customer (see customer signature below) and record all data accordingly :

- **Contact Name-** The contact person, especially when a contractor is dealing with large companies, usually corporations, shall be that person authorized to represent the customer. In the corporate world, it is usual for a contract to be signed by a corporate officer and the work overseen by an employee or an agent—the contact person. A contractor should be aware that his contact person, unless designated in writing by the customer, has no authority to sign any change orders or alter any terms of the contract. It is therefore prudent, in these cases, to request a written statement from the customer designating their authorized representative(s) and their limitations, if any.
- **Name-** The name of the company, corporation, or individuals should be written on this line.
- **Address, including zip code-** If the customer uses a Post Office Box, insist on an actual street address. This is needed for certified mail and other services. Trying to get this information later on when is needed it can be burdensome, expensive, time consuming and at times impossible.

3 **Proposed Contract Amount-** A decimal amount should be rounded off to the next full dollar. This method makes record keeping and billing easier.

4 **Payable as follows-** A contractor has to be specific on how he intends to get paid. In the clause, "We propose hereby to furnish labor and material as specified herein, for the sum of $_____ payable as follows," the most important phrase is "as specified herein," meaning no outside specifications apply to the price unless otherwise stated in the proposal.

- **Payment Schedules-** For service work contracts a contractor may choose a payment schedule that will include a down payment followed by two or three payments, as shown on the sample form. For those who are new to down payments, be assured that many good customers, especially in service work, are surprised when the contractor asks for none. For more on this subject, see "Negotiations" later in this manual.

- **For contracting-work-** contracts a contractor may use the same approach as in service work, or he may set a separate payment schedule, in which case he may include a clause that reads: "Payable in progressive payments as per contract breakdown to be submitted by the contractor prior to commencement of work." For more details on contract breakdown, see "Payment Requisitions" later in this manual.

5 **Proposal Withdrawal-** Every offer has a beginning and an end. In this case the proposal will end in 10 days, after which time is the contractor's choice to either accept or reject the contract. A contractor should consider the clause more of a customer motivator, to induce him or her to act quickly, than an offer deadline.

6 **The contractor signature-** Bear in mind that a Proposal / Contract is a ready-made form intended to remain a proposal until signed by the parties and, as with other contracts, both parties should sign it simultaneously unless special precautions are taken. However, when a contractor carelessly signs and leaves it behind as if it's a conventional quotation form, and waits for the customer's signature, he is making himself vulnerable. Because, regardless of how many copies he's holding back, he's leaving the final drafting of the contract to the discretion of the customer, who can readily modify and countersign it into the only binding contract copy—a feat that unfortunately is becoming popular with unscrupulous customers. Unless discovered immediately, it may not surface until the contractor is well into the job.

- **Unsigned Proposal / Contracts-** are limited offers proposed for a contract-to-be if accepted and executed within a given period of time. And if a contractor wishes to stay in tune with the industry by signing the contract before the customer does, he has a safe alternative: Submit the Proposal / Contract under a signed cover sheet or letterhead indicating his willingness to sign the contract as drafted within 10 days, or, when he feels he has to leave a signed Proposal/Contract for the customer to sign later on, the contractor should take the precaution of closely scrutinizing and comparing his retained copy to the one his customer sends back executed, before he invests any money in the job.

7 **Scope of Work-** A contractor should be careful in writing the scope of work—he should not get lazy and sacrifice thoroughness and clarity for easy-to-use general statements. For example, the ambiguities of the frequently used phrase *"furnish and install all electrical work as per plans and specifications"* without specific reference to the scope of work or to quantities has proven over and over to be costly to many contractors. Clarity is what a contractor wants in your contracts. Note the difference between "Install electrical wiring. . ." and the more popular phrase "Furnish and Install all electrical work. . ." The word "all" has no limit and *"electrical work"* includes all sorts of equipment such as magnetic starters and controlling devices that, most likely, a contractor has not included in his estimate.

In describing the scope of work, a contractor should describe it as "electrical wiring", for that is what he basically sells and does. As to quantities and true scope of work, a contractor should avoid using "all" whenever possible; he should use actual quantities and make reference to the architect's name, drawing numbers and dates as shown on the sample form. In absence of plans, a contractor should enumerate and describe each item of work clearly, using a logical approach beginning with location of work, followed by quantity and definition, as shown on the example form.

8 **Customer Acceptance-** Here is where a contractor outlines the terms under which he's willing to do business. He should study each clause and, if necessary, modify it to better serve his business. When he has finished, if need be, he should consult his lawyer and then draft his final business policy. Remember, this is the contractor's offer made on his own Proposal / Contract form and, as such, favors him. Therefore, in quoting and contracting work, it will always be to the contractor's advantage to more or less jump the gun and use his forms whenever possible, rather than stepping aside and leaving the advantage with the customer. In dealing with general contractors and most large accounts, a

contractor is expected to sign their contract documents which are more complex than the Proposal / Contract shown here (see Chapter 8 "The Contract"). The principal elements, however, remain the same. A contractor should read through the clutter of paragraphs and the true Solicitation, Offer, and Acceptance should readily become apparent. And while he is on this expedition, he should be certain his simple terms of doing business, as outlined in the list that follows, are not violated beyond acceptance:

- The price, specifications, and conditions set herein are satisfactory and hereby accepted. You are authorized to proceed with the work as specified herein.
- Payment shall be made as outlined above. Unpaid balances, after due date, are subject to 1.5% per month interest rate until paid in full. Customer agrees to pay all of the Contractor's costs related to the collection of any sum due, including legal fees and expenses.
- Patching, painting, site restoration, and the like generated by this work shall be done by others. Lamps, appliance cords and caps shall be supplied by others.
- Power company charges, Municipal permit fees, and Contractor's processing fees, if any, are not included in this proposal.
- The electrical permit shall cover only the work contracted herein. The removal of any existing violation is not included in this proposal. The Contractor, when required, for an additional charge to the Customer, shall apply for an electrical permit to cover said extra work.
- Changes to the scope of work shall be written as "Change Orders," and executed by both parties prior to commencing any work on said changes.
- Customer please note: When this Proposal / Contract exceeds $3000, in addition to the Contractor's signature, it requires the Contractor's corporate seal affixed where indicated above, or the contract shall be considered null and void.
- I (the customer's name printed) am authorized to accept and sign this contract because I am the Customer named above, or I am acting for the customer as his agent.

Chapter 4

SERVICE AND CONTRACTING WORK

BRANCHES OF WORK

The two fundamental branches of the electrical contracting business are: Service and Contracting Work. Each is a specialty in itself, requiring different technical and business approaches. For optimum results, companies engaged in both activities manage them under different departments.

Start-up companies, however, because Service Work is their springboard to Contracting Work, keep both activities under the same management. Manpower, tools, and administrative help are shared, requiring a keen managerial team capable of managing both activities.

In this chapter the business tools we are most concerned with are the commonly used business forms. And as with all other forms you'll encounter in this manual, don't be deceived by their familiar appearance and quit or skip reading. Study them in detail to achieve a full understanding of their value.

SERVICE WORK

The sample forms illustrated in this section, with the exception of the Service Call form, are intended to be used in Service Work as well in Contracting Work. As with all prepared documents, they also favor the writer—in this case the user.

The Service Call Form

A well laid-out form designated exclusively for service calls is the most important tool with which a contractor can equip his servicemen. The Service Call form, a stand-alone instrument, is unique because it progresses from quotation to contract to invoice and to release within the short period of time in which a service call is carried out. Through this form, a contractor will instruct his serviceman on a particular policy for a particular customer—for example, how he wants to get paid and how much to discount his list price.

Filling in all the blanks will greatly improve the field and office administrative performance, not to mention collection time.

The following comments address only those items pertinent to the understanding of the form as a whole:

John Doe Electric Co.
Licensed Electrical Contractor E-12345
123 Main St, City, USA, 12345, Fax (123) 456-7890

Service Call
Invoice No: _____
(123) 555-6400

Customer		Job	
Contact:		Contact:	
Name:		Name:	
Add:		Add:	
City/St/Zip:		Owner:	
Phone: Hm:	Wk:	Phone: Job:	Wk:

Date of Order	Order Taken By	Start Date	Cust. Order No.	Invoice Date

Terms		Paid	Serviceman No	Completion Date
COD ( ) Charge ( ) Cash Only ( )		Yes ( ) No ()		

Dispatched ____:____   Arrived ____:____   Out ___:___   In__:___   Out ___:___

Authorization to Enter the Premises and Commence Work

I authorize the contractor and his technicians to enter my premises (home/office/plant) to repair and/or install electrical work. I also promise to pay for all work performed in accordance with the terms and schedule rates shown below.

CUSTOMER SIGNATURE: _____   DATE: _____

Service Call: (Includes travel time to and from job site.)	Rates	Invoice
A. Regular time	$49	$ _____
B. Sunday and Holidays	$65	$ _____
C. Any day between 6:00 pm and 11:00 pm	$65	$ _____
D. Any day between 11:00 pm and 8:00 am	$85	$ _____

Labor Rates per each 1/2 hour unit or part thereof:

A. Regular time	$35 Ea x ___	units	=	$ _____
B. Sunday and Holidays	$45 Ea x ___	units	=	$ _____
C. Any day between 6:00 pm &11:00	$45 Ea x ___	units	=	$ _____
D. Any day between 11:00 pm & 8:00	$55 Ea x ___	units	=	$ _____
E. 3 hours and over quoted at	$ P/hr x ___	hours	=	$ _____
		Labor Total (a)		$ _____
	Material and Direct Job Expenses (b)			$ _____

Please Pay   Invoice Total (a + b) $ _____

Customer's Acceptance of Work

I hereby acknowledge the satisfactory completion of the work described herein. I agree to pay all of the contractor's costs related to the collection of any sums due, including legal fees and expenses. Patching, painting, and site restoration, if required, shall be done by others.

CUSTOMER SIGNATURE: _____ Title: _____ Date: _____

For C.O.D. please fill the following:
PRINTED Name: _____ Dr. Lic. # _____ State: ___ Exp. Date: _____

Work Description

_____
_____
_____

Form G-120-rev05 ©2005 The Estimating Room™ Inc.

Figure 2: Service Call

Comments on Service Call Form

1 **Authorization To Enter Premises And Commence Work-** This clause, aside from giving a contractor permission to enter and commence work, sets the foundations for a binding contract. The customer accepts his offer to do work at his scheduled rates and promises to pay in accordance with the contractor's terms and conditions.

2 **Dispatched . . . Etc.-** Here the contractor and the customer will agree and document the amount of time the serviceman spent for the job. Notice *'for the job'* and not *'at the job.'* *'For the job'* includes travel time to the job and to suppliers. This information must be entered prior to having the customer sign the *"Acceptance of Work"* clause. If there is any dispute, especially with charge account customers, the best time to resolve it is before the serviceman leaves the job, and not later on when the contractor is trying to collect his money and all is forgotten.

3 **Rates-** A contractor may refer to this column as his list price and offer a discount to his better customers which can be reflected in "This Invoice" column.

4 **This Invoice-** In this column the serviceman extends the charges earned by the company. They can be calculated at full rate or discounted as the contractor wishes.

5 **Labor Rate And Units Of Work-** In this form the labor rate is set at $35 per one 1/2 hour unit or portion thereof. Let's say the work was completed in 40 minutes; you're entitled to 2 units, or $70.

6 **Hourly-** For any service work that exceeds 3 hours, John Doe Electric will convert his charges to hourly rates. The hourly rate should be posted here prior to the customer signing the "Authorization To Commence Work" clause.

7 **Customer Acceptance Of Work-** Here the customer, by signing this clause, releases the contractor and accepts his collection terms. For cash customers paying by check, as a precautionary policy, a

good rule is to get the customer's Driver's License information, as well as the signature.

8 **Work Description-** Identify the work performed and any other work condition. For example, when the serviceman comes across a violation or hazardous condition, besides pointing it out to the customer, he should also write it down in the work description space.

The overall purpose of this form is clarity and protection, especially with first-time customers who wish to use a contractor's services. By putting in print the service call and labor rates, a contractor gives the customer the opportunity to accept or reject his service, thus avoiding misunderstanding and disputes later on. Bear in mind that most customers welcome this candid approach.

Nowadays, obtaining the customer's acceptance of work first is becoming the prevailing way of doing business.

THE WORK ORDER FORM

A work order form, as the title implies, is an order to carry out work on existing contracts, change orders, or just a day's work for a given customer. Its purpose is to clearly record the work performed in a format that will make billing easy for the office and enable the customer to review the charges. Each line is self-explanatory and, as with all other forms, filling out this form in its entirety will greatly improve the field crew work and the office billing and collection procedures.

Comments on Work Order Form

1 In this line we define the type of work order we're issuing:

- **Contract Work-** Checking this box indicates the work described is work of a contractual obligation, such as a punch list or repair work for an existing contract. When this box is checked, there shall be no charges to the customer.

- **Extra Work-** Checking this box indicates that the work described pertains to a specific change in an existing contract. Here you'll specify the Job No. and the approved Change Order No. The Ticket No. is kept in sequential order. For example, the crew works on C.O. No. 3 for Job No. 03-1210 for 3 days. If the customer wants the contractor to support all charges and the contractor wants his verifications approved at the end of each working day, he will fill out a Work Order for every day he worked. On each of the 3 forms the Job No. and C.O. No. will remain constant while the Ticket No. will be 1 for the first day, 2 for the second, and so on. This method will simplify bookkeeping and make backtracking charges easy.

- **Daywork-** Checking this box indicates a short-term job, from one to a few crew days, usually ordered by house-account customers. The work is done without written contract, billable on completion or monthly.

2 **Material and Direct Job Expenses-** Obviously, a contractor will post a lump sum amount here, and on a separate sheet, if need be, he'll list all material plus whatever markup he agreed to.

A Work Order, when used the way it's intended, is the contractor business' work horse. Backing up his charges accurately will bail him out of the most complex billing situations.

John Doe Electric Co.
Licensed Electrical Contractor E-12345
123 Main St, City, USA, 12345, Fax (123) 456-7890

Work Order
Invoice No: _____

(123) 555-6400

Customer	Job

Contact: Contact:

Name: Name:

Addr: Addr:

City/St/Zip: Owner:

Phone: Hm: W   Phone: Job: Wk:

Contract Work ( ) Extra Work ( ) for Job No. _____C.O. #___ Ticket #_____ Daywork ( )

Date of Order	Order Taken By	Starting Date	Cust. Order No.	Invoice Date

Terms	Paid	Serviceman	No.	Completion Date

COD ( )   Charge ()   Cash Only ( )   Yes ( )   No ( )

Dispatched ____:____   Arrived ____:____   Out ___:__   In ___:____   Out ____:____

Date	Day of Week	Workman / Crew	Hrs Worked	Hourly Rate	Amount
				$	$
				$	$
				$	$
				$	$
				$	$
				$	$
				Labor Total	$
			Material and Direct Job Expenses		$
				Invoice Total	$

Please Pay

Work Description

(This and other forms may be downloaded online at www.theestimatingroom.com)

Customer Acceptance of Work

I hereby acknowledge the satisfactory completion of the work described herein. I agree to pay all of the contractor's costs related to the collection of any sums due, including legal fees and expenses. Patching, painting, and site restoration, when required, shall be done by others. By signing this document, I accept the terms and conditions set herein.

CUSTOMER SIGNATURE: _____ Title: _____ Date: _____

PRINTED Name: _____ Dr. Lic#_____ State:___ Exp. Date:_____

Form EG-110-rev05 ©2005 The Estimating Room™ Inc.

Figure 3: Work Order Form

THE CHANGE ORDER FORM

Any job a contractor undertakes, no matter how well it was designed, is bound to change. When that occurs, as stipulated in most contracts, before a contractor proceeds with any changes to his scope of work, he must first obtain a written authorization from his customer. This authorization is well served if written in the form of a Change Order, wherein pertinent information is documented and agreed upon among the parties. (See Chapter 11, "Contract Management.")

Notes to the Change Order Form

1 **Contract No. and Date-** A contractor should fill in the job's original contract number and its date. This information is needed to tie the original contract terms and conditions to its change orders.

2 **Terms-** According to job condition and customer relationship, it's conceivable that a contractor wants to get paid in cash for any extra work; therefore, the Change Order form provides him with Cash or Charge conditions. In small jobs, especially private work with down payments, it is good business to get paid as additional work is performed.

3 *Documentation-* Complex or extended changes most often require supporting documentation. On this line a contractor will have an opportunity to record the type of documentation he's submitting with his change order, if any.

4 **Extension of time-** For any change to the scope of work, there is always a period of time that is required to perform the task. In the space provided, a contractor can enter the numbers of days or, in case of small changes, the fractions of a day he wants his contract extended by. For example ¼, ½, or ¾ of a day. A contractor should not underestimate its value. At times, especially toward the end of the contract, he'll wish he had recorded and requested extensions as he went along.

5 **Customer Acceptance-** It is the tendency of unauthorized personnel, especially in large jobs, to fill or want to fill the shoes of those who are authorized to execute documents on behalf of the customer. Here, by the nature of the change order form, a contractor is able to question them and commit them to their claim

in writing. When he's in doubt of their authority, he should check their status with their superior. On this subject an important note to mention is that many unscrupulous customers may lead a contractor to believe their field representative is authorized to sign changes only to negate them when it comes time to collect. It is therefore the intent of this change order to discourage the unauthorized individual from imposing himself or force him to corroborate with the customer.

John Doe Electric Co.
Licensed Electrical Contractor E-12345
123 Main St, City, USA, 12345, Fax (123) 456-7890

Change Order
No: ___
Date:___/___/___

Customer	Job	1

Contact:

Name:

Add,

City/St/Zip:

Home: Wk:

Beeper: Fax:

2

Job Name:

Contract No.   Dated:___/___/

Add.

Owner:

Job Phone:

Terms   Paid   Charge To

C.O.D. ( )   Cash Only ( )   Yes ( )   No ( )   Cust/GC: ( )   Others:

3

Documentation

No ( )   Not Applicable ( )   Will Follow ( )   Yes ( )   No of Sheets:

The Contractor and the Customer agree to make change(s) as specified below for: $

Previous Contract Amount including all previous Change Orders: $

Revised Contract Amount: $

4

With this Change Order, the contractor requests _____ day(s) extension to the project completion date. No work shall commence on the work covered by this Change Order until both parties execute it.

Contractor's Rep: _____Signature: _____Date: _____

Changes To Original Contract

(This and other forms may be downloaded online at www.theestimatingroom.com)

Customer Acceptance

The price and conditions set forth are satisfactory and are hereby accepted. The contractor is not responsible for redesigning or upgrading any existing power distribution system that might be affected by the added electrical load of this change. This Change Order, upon its execution, becomes part of and conforms to the terms and conditions of the existing contract as identified above.

I (Print name) _____ am authorized to accept and sign this Change Order because I am the customer named above, or I am acting for the customer as his agent.

Customer Signature: _____Title: _____Date: _____

Form EG-130 ©2005 The Estimating Room™ Inc.

Figure 4: Change Order Form

CONTRACTING WORK

So far in doing service work, a contractor has been in control of most transactions. He has been able to use his own forms. In contracting work, however, as he begins to work for general contractors, home improvement companies, property managers, and the like, he'll discover that his standard forms have less standing. For example, most prospective customers use their own contract forms. If a contractor wants their jobs, he will have to use their forms.

While this is true in most cases (and if a contractor remembers that written documents always favor the writer), he should always present his proposal on his own quotation or proposal / contract form, even when he thinks it will not be used as the final contract document. There are two basic reasons for doing this:

(1) A contractor should never assume that the customer, even if he's a general contractor, will not sign his proposal. If his proposal, following this manual instructions, is well written, the contractor should not be surprised if the customer seizes the opportunity and, either to save paperwork or to expedite the job, signs his contract, on the spot, just the way he submitted it; and

(2) In case of conflict between his proposal and the final contract, a contractor can readily make reference to his proposal and demonstrate his initial intent. In fact, when a contractor signs others' contracts, besides reading and comparing them carefully to his original proposal, he should always attempt to make his written proposal, either by reference or by exhibit, part of the final contract document.

In contracting there are two questionable practices, mostly in private work, which a contractor must learn to deal with, for they are vastly accepted as ways of doing business and are not likely to change in the near future.

The first questionable practice is to bid a job based on a limited set of electrical drawings supplied by the customer, so that later on, when the job is awarded, a contractor is expected to sign a contract listing, as part of his scope of work, an array of additional drawings and specifications he didn't know existed. The second questionable practice is accepting a verbal contract even though the contractor doesn't know the content of the customer's contract documents along with their terms and conditions until he receives them in the mail, usually a day or two before or after he started the job.

The evolution of these practices, or customs if you will, stems from the customer's sometimes intentional procrastination in preparing the final contract document, and the contractor's fear of losing the job if he rocks the boat with premature demands. Regardless of the reason, both practices can create problems because if there is no mutual agreement, both sides stand to lose.

To prevent falling victim to these questionable practices, a contractor must refrain from building up undue expectations, even when the job is committed to him by a letter of intent. Until he has reached and signed an agreement with his customer, a contractor has no contract and therefore no business to proclaim the job in-house and make all sorts of irrevocable commitments, such as pulling permits and ordering material and equipment. Refraining from making these kinds of commitments is crucial in protecting his interests as well as those of his teammates.

Ironically, as long as contractors fail to review the entire contract document and, when necessary, take precautions before they decide on bidding a job, a majority of unfair and lopsided contracts will continue to be signed, with most of them ending up in litigation.

To properly bid a job, a contractor must know the full scope of work, including the customer's lopsided terms and conditions, for only then can he decide whether or not to bid the job, and if so, include the extra cost in his estimate.

In contracting, what will hurt a contractor most are the unknown and hidden costs. Therefore, the road to successfully completing a job, as he will learn, is very narrow and usually left in the hands and opinions of others. For sample contracts and discussion on this subject, see Chapter 8, "The Contract."

Perspective
When you can assume the viewpoint of the business as a whole, rather than just the viewpoint of an individual worker, you become immensely more valuable and productive. The material in this chapter will help you understand how the business side of the electrical industry works by taking you into the mind of your employer, the electrical contractor.

Chapter 5

THE STARTING POINT

WHERE IT ALL STARTS

It all starts with a sales lead. Regardless of the marketing methods employed by a contractor in promoting his business, the result will always be a lead to potential work.

Each sales lead is a new starting point whose outcome is totally dependent on the contractor's decision to accept or reject it. In making that decision, he must bear in mind that unless he's ready to carry it out to its conclusion, he should not accept it. Accepting a sales lead is a commitment to invest his resources, not to mention his reputation as a reliable contractor, into something whose cost gets progressively larger with the extent of his involvement.

Taking on a sale lead is the start of a new business transaction, a "loop" if you will, which like many business transactions, is worthless unless it has the potential to close into a good sale. If, at any time during the course of the loop, the lead's potential is poor or declines, the contractor should not hesitate; he should quickly close the loop and walk away. If he hesitates, he will not only increase his cost with every passing moment, but will also impose undue disappointment on those he's dealing with. For example, abandoning a sales lead three quarters of the way into its loop, when he could have done so at the start, is wasteful and distracting to all parties concerned, and that includes his own staff.

Therefore, for a contractor to get his business on the road to success there is no better way than to evaluate each sales lead before he takes it on. Sales leads are the roots of his business. A lead develops into a job, and a job into a customer. The customers' quality determines his company's quality, and thus his success rate.

The knowledge he'll gain on the many aspects of prospective customers and job procedures will assist him in evaluating sales leads, so he can invest his resources (time, energy, and money) on productive business, rather then on chasing wasteful dead-end prospects.

Before undertaking the task of following up a sales lead, regardless of whether it's for a new or an old customer, a contractor should routinely go through an inquiry checklist, with the following questions being at the top of that list:

- How will I get paid?
- Who is the customer?
- Can I successfully deal with this customer?
- Do I have the finances and organization to deliver a successful job?
- Who is my competition?
- Can I be competitive and still make a profit?
- If all answers are favorable and he decides to take on the sales lead, a contractor should use his best resources and go at it wholeheartedly, but if any answer is unfavorable, then, no matter how attractive the lead is or whatever justifiable reasons he comes up with, he should not second-guess himself, he should just walk away. If it hurts him, it will only be for a couple of days or until he has found a better lead to invest his resources on.

VULNERABILITY

To reinforce the importance of evaluating sales leads, here are a few compelling thoughts on the subject of vulnerability.

The ranks of contractors include individuals whose optimism goes beyond the norm. As many of them confess, they believe things will get better with each devastating blow they get. For example, if the building collapses, they'll make more money from the insurance claim; and, if the scope of work is changed, they'll make more money from the extra work and so on. At least, that is the way they think.

Along with this trend of thought, these individuals, no matter how many times they get hit, always rebound thinking that the next customer is their best friend with their best interests at heart, ready to pay in full for every job they do. In essence, these individuals have the ability to put aside bad experiences, trust newcomers, and treat every new job as a new start. The only difference between this kind of contractor and the wiser contractors is that these contractors, like gamblers, will have to start from scratch every time they lose until they learn how to operate from facts rather than feelings.

On any given day, anyone can call on this contractor to quote a job, and most likely he will invest his last dollar on doing the job without ever questioning the customer's ability to pay. He will work on the assumption that he has a standard contract signed, whatever that means, and that he will get paid in full, especially if he receives one or two partial payments in the interim.

If you think this is exaggerated or fiction, think again. Every day it happens countless times in our industry. If you never experienced it, the following short story, perhaps, will enlighten you:

A few days into demolition, a construction manager for a national chain restaurant went through the yellow pages and invited various trade contractors to bid his work. Those interested were asked to attend a pre-construction meeting where they could visit the job and examine the bid documents.

At the meeting, the attendees—plumbers, electricians, mechanical contractors and the like—as masters of their trade, each got promising answers to their technical and other questions pertaining to schedules, scope of work, material, specialty items, etc., and after almost a two-hour session they could feel the excitement in the air. Each tradesman, if asked, was ready to start building right then and there.

What makes this typical story compelling is the fact that not one single contractor asked the most important question: "How do I get paid?" and when a wiser contractor finally asked, neither the construction manager nor the owner's representative could readily answer the question. As to the contractors present, most of them thought it were an impertinent question, as payment schedules are standard procedures, and with that question a complete silence set in the room, as if a cardinal rule had been broken. However, when asked again, the construction manager had to call the owner and ask which brought the group to the harsh realization that neither the owner's representative nor the construction manager had given much thought to payment schedules.

Nevertheless, the restaurant was built, regardless of payment conditions, by those who failed to heed the warning and who most likely are still hoping to collect on their final payment.

This is not to insinuate that owners or construction managers go around having jobs built with no intention to pay, even though our carelessness in this regard at times is the biggest temptation to them of all. This is merely to remind you how easily enthusiasm can take a contractor into bad investments.

Going back to our short story, either by rule of thumb or actual checking up, each contractor in that room could have avoided any problems. Those that declined the invitation suffered no losses and went on to better jobs. The optimists are still waiting for their payments, not to mention their profits.

When a contractor attends pre-construction meetings, aside from asking vital business questions, he should pay special attention to the other tradesmen. Their line of questioning can tell him a great deal about them and the type of cooperation he can expect on the job. The expertise and cooperation of other trades in carrying out their work diligently greatly affects the outcome of his work. For example, a bad masonry or mechanical contractor can upset the progress of the job and drastically decrease his productivity.

Checking the other contractors' reputation is part of a contractor's prospective job assessment. When he's asked to bid work, he should not hesitate to ask who the major trades are. It may sound unorthodox, but he should make a few phone calls if he has to, for it is all part of his investigative procedure to prevent problems and assure success in all the jobs he does.

WHO'S YOUR CUSTOMER?

The customer is that person who purchases goods or services from a contractor. If a contractor does a job at City Hall for a given general contractor, his customer is the general contractor, not City Hall. His customer is that individual or firm that hires him to do the work and with whom he has a contract regardless of whether it's written, verbal, or implied. This is a relationship that a contractor should never lose sight of in all his dealings. For example, in the case cited above, if he wants to maintain a legal, moral, and ethical standing, he cannot take orders from City Hall, nor should he bypass the general contractor and deal directly with City Hall.

General Contractor

A customer, therefore, can be anyone with whom a contractor deals directly, from a homeowner to a property manager to a general contractor. In contracting, the most prominent customer he's bound to deal with is the general contractor, and therefore a contractor should learn all he can about his character.

Workable Relationships

The general contractor-subcontractor relationship has to be straightforward, for both parties have a common goal—making a profit in the construction of the project they contract. However, to earn these profits they depend on each other's abilities to carry out the work. Each must do what he does best, that is, use the tools of his trade in the most profitable and professional way. For our purposes we shall refer to this kind of relationship as a "working relationship."

Understanding this relationship and the relative positions to each other and to the project is what makes winning teams. The general contractor must understand the subcontractor's position, and the subcontractor must understand the general contractor's position. Neither should feel threatened by the other's actions when dispatching their duties, especially when they fall within the contract's legal framework and the industry's standard practices.

As the title implies, the general contractor is the individual or firm who contracts the building of a complete project, most likely on a turnkey basis, and at a fixed cost. His position is relatively risky and demands the respect and support of his subcontractors.

Due to our industry's inherent competition, a general contractor, in order to secure a contract, quite often makes extraordinary concessions to the owner. In a working relationship, subcontractors, when asked, generally will share the burden of these concessions. For contractors are bold people who do not fear bad weather, earthquakes, or the like, nor do they anticipate unforeseen events that can jeopardize the completion of a project on time or upset the anticipated construction cost. They can always do better. (And a good thing too, or no building would have ever been built).

The subcontractor, such as the electrical contractor, is that individual or firm who carries out a portion of the general contractor's work, and in using managerial skills, attempts to make the most profit he can.

To reach his goal, however, he must recognize that a good working relationship with the general contractor is vital. For example, if he wants to be adequately compensated for extras or obtain an approval on a substitution for an item of work, he cannot and should not expect the general contractor to prepare his submissions or do the legwork that is often associated with those requests.

In seeking approvals for substitutions, perhaps the contractor and his supplier are the experts and only parties interested in its outcome. And, if the outcome is important to him, then no one but himself should handle the details and presentations required by most submittals.

With regard to additions or deletions of work, a contractor should submit change orders on time. Change orders should be complete, including explanatory notes detailing the reason for change and estimate breakdowns justifying the cost.

This clear method will empower and encourage his teammate, the general contractor, to go to battle with the owner not only for the contractor, but for himself; for he has given him tools with which he too can profit.

Anything short of these methods will only jeopardize whatever profits a contractor is entitled to make on these change orders.

To approach these procedures differently—for example, to merely ask the field superintendent to process them—would probably result in a forgotten promise. This can create a real disappointment later on when the contractor finally realizes that he's not getting paid for the extra work and he has lost all basis for a timely and legitimate claim.

As a subcontractor, a contractor must assume the responsibility to document his work properly and submit what has to be submitted to the general contractor in a timely fashion. There is no room for procrastination in these transactions, especially with the ones that have to do with protecting his own interest—his business.

Basically, this is the respect and relationship that should and must exist between general contractor and subcontractor. Often, regardless of needs and conditions, some relationships are best not started. There are no plausible excuses for knowingly entering a lopsided relationship. If for no other reason, a contractor should be selective of the general contractor he chooses to work with because he will have a great influence over the profit a contractor makes on his investment.

The subordination to the general contractor presumably is motivated by the contractor's optimistic view of the greater number of opportunities he has to generate a higher percentage of profit than the general contractor.

But, whatever the view or motive is, a contractor must not lose sight of the general contractor's direct control over moneys due to him. In most cases, the general contractor has the upper hand and can and will cut into the contractor's profit by back-charging whatever expenses he thinks the contractor owes him—the core of much litigation, many disappointments, and many failures.

Perhaps here, in this unbalanced formula of equity and anticipated profits, lies the mystery of what makes most of these relationships work. Leadership and subordination perhaps could be additional reasons. Regardless of the answer, a contractor must recognize a bad relationship before it starts and prevent it by following his intuition and looking for work elsewhere.

Non-Workable Relationships

Non-workable relationships, for our purpose, are those relationships that, quite often, make us question our judgment and make our discontent grow to unbearable levels. Unworkable relationships also have the tendency to make bank accounts plunge toward dangerous balances.

In theory, however, if a contractor's estimates are the true reflection of anticipated costs and he is able to carry out the work contracted, then the job can be classified as a good job. And if he and the general contractor pay steadfast attention to what is agreed on, then the relationship can also be classified good and, therefore, workable.

Simply put, that is the theoretical aspect of a good job and a workable relationship. Workable, that is, until the unscrupulous human element takes over the relationship.

The introduction of this human element into the equation requires the contractor to be alert. In today's business, the following two elements concern the contractor most:

The Unscrupulous General Contractor

The unscrupulous general contractor is the first to come to mind. Sooner or later he will cross the contractor's path. And when he does, his actions will leave him with a long-lasting negative impression.

His unscrupulous behavior toward the contractor is frequently encouraged by the contractor's seemingly vulnerable financial position.

To assist the contractor in identifying, and possibly preventing, some of the most unscrupulous contracting methods, the following list has been compiled. While we are aware that, occasionally, some of these practices come about inadvertently, nevertheless we must recognize them here:

- Requesting a quote on an incomplete set of drawings.
- Withholding bidding documents from the bidding package.
- The usage of lopsided contracts, instead of standard industry-accepted contracts.
- The showing of others' bids to lure the contractor into lowering his price.
- Withholding full 10% retainer on payment when theirs was reduced to 5% at the midpoint of the project.
- Unduly cutting down the contractor's monthly requisition.
- Evading and delaying the issuance of bona fide directives and change orders.
- Improper documents signing.
- Withholding items of work, so as to hold up final inspections that will entitle the sub-contractor to collect his moneys.
- Refusal of certified mail.
- Some other items of concern to us affect job site productivity. These items arise from poor field coordination and a general contractor's procedural negligence:
- Unnecessary cluttering of floor space with premature delivery of building materials or fixed fixtures that hinder free movement of rolling scaffolds and similar equipment. This is done merely to save the general contractor on some rental storage trailer.
- Improper field supervision and coordination of trades.

Unscrupulous Plans and Specifications Writers

Conflicts between plans, specifications, and contract documents are usually accidental but still costly when they go undetected during the estimating stages of a project. However, accidental conflicts that are generated by designers, specification writers, and project managers due to their lack of knowledge or responsibility should also be classified as unscrupulous contracting. (For more on this subject and how to prevent it, see Chapter 10, "Protecting the Job.")

Perspective

When you can assume the viewpoint of the business as a whole, rather than just the viewpoint of an individual worker, you become immensely more valuable and productive. The material in this chapter will help you understand how the business side of the electrical industry works by taking you into the mind of your employer, the electrical contractor.

Chapter 6

WHERE TO LOOK FOR WORK

FORESIGHT

This chapter is about how to find work for both the beginning and the seasoned contractor. If a contractor has read and subscribed to the methods introduced in the previous chapters, he should be ready and anxious to apply his newly acquired marketing skills, eager to pick and choose customers and jobs, and, when conditions don't fit his needs, he should have the conviction to walk away; for that's the way he should feel—eager and confident.

However, for a contractor to live by his convictions he must look at his market opportunities overview—the full picture—and feel comfortable with its prospective outcome. When he can see that, not only will he be able to live by his convictions, but he'll become a focused contractor with only one purpose—success.

It is this foresight that will help a contractor look for and find the work that best fits his marketing plans. If his sight is restricted, as it is many times, by limited knowledge of what's around him or by self-imposed limitations, then he's doomed to fail, or at best struggle along just making ends meet.

If a contractor needs a clear demonstration of what we are talking about in this chapter, he can get it by making a very small investment. On a clear night he should take a short helicopter ride over his city, and suddenly he'll see and feel with a sense of brilliancy all the prospective work awaiting him. He'll see the street lighting, the sports fields, the cinemas and malls and shopping centers and the sewer treatment plant and the substations and switching stations, the hospitals, the churches, the government center, and the empty lots people will build on. He'll see the extent of his prospective work and, as advocated in this manual, the reason to be choosy, for he has the talent to sell and the know-how to repair, maintain, alter and install all that he sees.

It is with this panoramic sight in mind that we present the following methods and approaches to help the contractor pick and choose his prospective customers and jobs. A contractor can always come back to this list of opportunities to keep his marketing goals sharp and vivid in his mind.

In the following sections we shall cover the many market approaches to Service and Contracting Work.

SERVICE WORK

Service work, as defined here, is the most common type of work for commercial and residential accounts, and it is the most sought out work by start-up contractors. Service work consists of repair, addition, alteration and maintenance of existing installations. This type of work can be generated either by the private sector (for example, Realtors, property managers, home and business owners) or by governmental agencies, such as federal, state, county and municipal. Regardless of what category a contractor chooses, he must always remember that repeat business comes from loyal customers (accounts), and they can only be built by word-of-mouth recommendations. Each satisfied customer will earn the contractor that recommendation, which is the foundation of his service work.

THE ESTABLISHMENT

In recent years both private and public sector service work producers for electrical contractors have been switching from total patronage of service companies to a three-bids system for all work other than emergency work. While this method creates more opportunities in which the contractor may compete, it also imposes on the servicemen's daily schedule. For example, unless a contractor screens and evaluates his sales leads, as discussed in the previous chapter, he'll be giving bids all day long with little success.

In drumming up work, another thing to take into consideration is that no matter how much a customer likes the contractor, or how much work the contractor has done for his company, the award of the next job is always predicated on price first. In other words, a contractor must possess all the qualities a customer will look for in a reliable contractor, with price being always at the top of the list.

PRIVATE SECTOR

While the different methods of generating work from the private sector are generally simple, they can seem unorthodox to some. Nevertheless, the methods introduced here are successful for those that apply them.

To generate service work in the private sector, the contractor's approach is different than in contracting. For example, a contractor may knock on doors at an industrial park, or select any or all the following methods:

Directories

Directories such as Yellow Pages, Community Industrial Listings, and the like, when properly explored, are comparable to the helicopter ride mentioned earlier: through their pages a contractor may shop for his business markets, if you will. For example, If he's a one-man show, just starting out, and wants to create an immediate clientele with good cash flow, choose retailers of lighting fixtures, paddle fans, heat pumps, exhaust fans, spas, major appliances, and he should personally call on them for their electrical needs on a regular basis. If he's beyond starting out, and wishes to expand his service business, then choose home improvement contractors, interior decorators, heating and air conditioning contractors, etc. If he's a well-established contractor with a service work quota to fill, then he should get his sales forces to shop through the industrial pages for suppliers of specialty equipment for offices and businesses such as automotive spray booths, electric car lifters, electronic systems, moveable partitions installers, network computer specialists—and don't forget the insurance adjuster for flood and fire damages—and offer them his electrical expertise services.

In essence, if a contractor is in business, he has to use his imagination and exploit these directories to his advantage as they contain a wealth of markets for the electrical contractor. They are the most informative and inexpensive way to either start up or boost your business, especially when his marketing skills are applied with enthusiasm and determination, for they will pyramid into endless opportunities.

Yellow Pages and Local Directories

In some areas such as the American suburbs, ads placed in Yellow Pages and local directories do not generate as much service work as they used to before the do-it-yourself concept was popularized by Home Depot and building-supply stores. The cost of non-productive ads, especially in long-term commitments, can well overrun small companies' budgets.

After a contractor has taken into consideration the pros and cons and he feels he must advertise in these directories, he should do so on a trial basis. To do this, use a dedicated telephone number for each locality in which you wish to advertise. This method, which is encouraged by many publishers of telephone directories, will allow the contractor to cancel his monthly charges as soon as he shuts off the telephone number used in the ad.

A contractor should not use his advertising telephone number for any other purpose than advertising. His business cards and stationery should have his company's operating telephone numbers, thus directing his new customers to communicate with his office over those lines. Later on, if he decides to shut down the ad he'll be able to retain whatever customers he paid for.

Other hints to consider when placing these ads:

- **Size of Ad-** Larger or largest is not necessarily better. A contractor should look into four or five back issues (the library has them) of the directory he wants to advertise in and check how many ads of the size he's contemplating have survived more than two or three issues. Most likely, if the ad is three quarters of a page or larger, economics will restrict it to no more than two years.

- **Stay With the Crowd-** In any one directory there always is the largest, the smallest, and the in-between ad. The last is where the crowd is most likely to shop. Consumer habits will force the shopper to flip to the next pages for more choices. Page one with only one ad has the consumer cornered and scared of price gouging to pay for the large ad. After flipping over the page, the consumer will tend to stop in the middle, where there are the most choices. If the middle has one-quarter-page ads, he or she will have eight choices and will totally forget page one and all others in between, no matter what their messages are.

- **The Ad Layout-** It must be kept simple and free of clutter. For cost-effective ads, the contractor should review the specifications in Chapter 2, and use them as guidelines here. And use words that sell.

- **The Calls-** A contractor must remember these ads will only generate telephone calls; it is up to him or his staff to be there to receive them and turn them into jobs.

Knocking On Doors

This method can begin at the contractor's next-door neighbor's house and end at the door steps of the largest corporations. It all depends on how ambitious he is and how far he's willing to travel. Regardless of his goal, this method has passed the test of time. In a sense, a contractor may call it business politics at its best. He simply mingles with and lobbies, if you will, those who are most apt to give him work. For a list of who's who, their contact names, mailing addresses, telephone numbers and all pertinent information, he may subscribe to several specialized publications. The companies a contractor will most concentrate on in this endeavor are national chain stores, restaurants, fast-food stations, banks, department stores, and the like that are still entertaining the concept of service contracts for their outlets.

PUBLIC SECTOR

No electrical contractor should be in business without seeking a share of public work. His participation in public works, especially service work, is as essential as joining the local chamber of commerce, or any electrical association. Doing work for the local municipal agencies, besides enhancing the contractor's image as a reliable contractor, generates steady work for his company—sort of a security blanket for when the private construction market slows down. These are markets a contractor must always cultivate, for they are the bedrock of his business.

It is a common temptation for most contractors, however, as soon as they get busy with private work, to abandon all sales efforts for public work. This demonstrates that these contractors have not evaluated their marketing plans, or the consequences of quitting.

Many public agencies keep a list of bidders to whom invitations-to-bid documents are sent. For a contractor to maintain his position on these lists, as most of them are extensive, when he's invited he must be responsive. If not, after a period of time, usually a year, or a certain number of no-responses, he'll be dropped from the list—an event that will not favor him the next time he's looking for work with that agency.

The process of becoming an accepted contractor, from the filing of applications until he is regularly invited to bid work by the various government agencies (aside from the so-called lottery or rotation system you'll hear about) will take some doing on his part, requiring perseverance and performance. As in the private sector, when doing service work for public agencies, a contractor must build a solid reputation for dependability. If he pays close attention to their needs and complete the work on time and as specified, contract administrators will naturally begin to favor him in their next round of invitations. In other words, he'll be pulled out of the list of prospective bidders and added to the list of responsive and dependable contractors. When that occurs, he will have closed a business loop successfully, for work will be looking for him, instead of him looking for work.

A contractor should remember, this is service work he is looking for in the public sector, where most agencies are authorized to award work without competitive bidding, up to a certain amount. A contractor should find out what those limits are with each agency for which he intends to work. To apply, a contractor need not visit any of these agencies. He can begin by calling the purchasing agent or the contracting administrator of the specific agency he's interested in, and have them send him the necessary documents to place him on their bidder's list. A contractor should study the material he receives from each agency. Each operates under a different set of rules and regulations.

Public Agencies

For a complete list of public agencies within his locality, a contractor may start with the white pages of the telephone directories and end at the Small Business Administration office. The following lists are meant to steer him in the right direction. A contractor should investigate any other areas or agencies that might interest him.

Government Agencies

Government agencies that are most apt to provide service work are:
- General Service Administration (GSA). Manages all properties owned and leased by the US Government such as post offices, office buildings, etc.
- Federal Aviation Administration (FAA). Responsible for all navigation and landing systems for private airports, including structures that house said systems.
- US Army Corps of Engineers

- US Navy
- Housing and Urban Development (HUD)

State Agencies

- Department of Transportation
- General Service Administration
- Department of State

County Agencies

- Board of Education
- Public Library
- Treatment Plants
- Hospitals

Municipal Agencies

- Public Works
- Recreational Facilities
- Housing Authorities
- Utilities

CONTRACTING WORK

Construction Reports

On a given day all sorts of prospective work can be funneled through construction reports right to the contractor's doorstep, or, thanks to electronic media, to his desktop computer without his traveling a single mile. Through these reports a contractor is able to feel the pulse of his industry in any locality he wishes. The frequency and age of the information he chooses to receive comes with a price tag that varies with style, territory, and immediacy. However, none of these services, when properly fitted to his business budget, will ever exceed the cost of sending his sales forces into the field.

A contractor can be informed on a daily or weekly basis of who's planning and designing what, and who's building where and when, or who won and lost yesterday's bids, with a complete list of bidders, their telephone and facsimile (fax) numbers, addresses and other pertinent information he'll need to bid the job he chooses.

While these reports are available to the contractor through private companies as well as through various governmental agencies, their final product is the same. They produce sales leads which a contractor must investigate before he puts his wheels in motion, as he would any other sales lead.

SALES APPROACH

In contracting, cold calls are seldom productive. In contracting, cold calls made at the wrong time of the day are never productive. The most effective time to contact a prospective customer, especially a general contractor who's on the go, traditionally is between 7:00 and 10:00 a.m. That is the time when we all think most clearly and are most receptive to doing business. In our industry that time of the day is when most deals are made, for the rest of the day is spent on job management and negotiations with those we called early in the morning. For example, knocking at a general contractor's door at two in the afternoon on a fishing expedition for work, without even a job name to allow the contractor's foot-in-the-door, will greatly decrease his odds, for this is a cold and wasteful sales call.

When looking for work, a contractor should take two things into consideration: (1) The time of day he's making the sales call, and (2) the knowledge of the job he's calling about. The second of these is a vital element of his sales call. The more knowledgeable he is about the job he seeks, the more confident his prospective customer will feel about him, and the better his chances of winning the job. At first, a contractor may acquire part of that knowledge through construction reports, and later complement it through his estimating process.

In summary, a contractor should look for contracting work through whatever means he chooses, then follow its progress from designing to bidding stages, attain the list of bidders and find the right time to contact those he wishes to work with. Once he has gotten his commitments, then enter the race by estimating and bidding on the job. This simple process depends on the contractor's sales lead analysis and is the reason why he should never call on a general contractor without a specific project in mind in which both of them have a common interest.

Contracting With Public Agencies

In some areas, mechanical trades for public jobs are bid through general contractors; in others, directly to the owner. However, no matter what system is adopted in an area, the basic bidding principle remains the same. Therefore, when a contractor is looking for work, if he does not wish to subscribe to paid reports and bid for work that is directly bid to the owner; then a contractor may want to apply to the various governmental agencies of his choice and ask to be placed on their bidders' mailing lists. In this case he should follow the same guidelines laid out earlier under the Service Work section.

When to Look For Work

A contractor should look for work when he is busy. He should not fall for the "I am too busy to look for work," syndrome, nor should he get distracted from sales by the first job he gets. Eventually he will complete it, and he'll be looking for work again, this time under pressure, and he will have lost his marketing momentum.

Chapter 7

THE BID PRICE

PRESENTATION

Once a bid is compiled, it would be counterproductive to lose a job due to a weak presentation. Weak presentations are the hallmark of those who lack confidence in their estimates and consequently in their bid prices, thus reflecting this weakness all the way into their proposal and presentation. The fact that a contractor burned the midnight candle or he knows the customer well, does not exonerate him from presenting a businesslike proposal.

A clear-cut proposal (see Chapter 3, "Basics of Contract, Notes to Proposal/Contract Form") is the hallmark of a successful businessperson who radiates confidence. Confidence is, of course, the foundation of a sound business relationship. Remember, most of us would like to know, in the least number of words, exactly what we're getting for our money. When certainty is lacking, so is confidence. For example, *"The work will cost you between $2000 and $3000"* is a sure loser to *"The work will cost you $2700 with the following exceptions"* complemented by a clear list of exceptions. Even if the first statement is well meant and the job's ultimate cost may be less than $2700, the second statement wins because it delivers confidence. The statement satisfies the customer's objective—

firmly knowing where he or she stands beforehand.

When a contractor is invited to quote a job, people want firm prices. If he's not sure about the job and needs to add some contingencies into his price, he should do it with confidence, regardless of consequences. When a contractor shows uncertainty he's telling the customer that he doesn't know what he's doing. While this is an innate syndrome with doubtful individuals, it is amazingly common, and at times habitual, among well intentioned contractors. Every now and then it just creeps up on them to kill a few good jobs. A contractor should review his quotation pattern, and if he's guilty he should take corrective actions.

PRICING STRUCTURE

The oldest and hardest barrier to overcome is the psychological effect the less-than-a-dollar pricing structure has over the consumer. For instance, $1.99 is not $2.00, just as $99,999.99 is not $100,000.00. If a contractor analyzes this principle and adapts it to his quotations and change orders pricing, it can make a great deal of difference in his business bottom line, especially if he's an average contractor doing lots of small jobs.

When pricing jobs, a contractor wants his number to deliver the most favorable impact while getting as many dollars as he can without jeopardizing the chances of getting the job. Because there has to be a number at which a contractor must freeze his bid price, cutting it may well give him the clearest and highest price he can get.

If the bid, for example, adds up to $10,050.25, the contractor should make it clearer by rounding it off to $10,050. In lump-sum bids, he should always round off his price to the next full dollar. Now, though he has made it easier to read, it's still too high and will not favor him. According to our marketing principle, the number is exactly $51 too high—an amount most contractors will gladly shave off their bid price just to get the job. The bid should be $9,999 or, if he feels funny about all the nines, $9,990. No matter how he looks at it, psychologically his new bid price appears to be far cheaper than it really is and less cumbersome than the original $10,050.25.

By doing this, he should break the barrier between $10,000 and $9,000, where the in-between numbers from 9,001 to 9,999 or from 10,001 to 10,999 have little or no effect on the initial price-impact. For example, unless he's engaged in fierce competition, bidding the job at $9,250 or $9,450 or $9,750 or $9,999 or any other number in the $9,000 range will not take him out of the customer's $9,000 cost-perception. The $9,000 becomes the leading number in his bid and any other bid above his, even by a dollar, will fall into the next plateau, the $10,000 cost-perception. And if, as advocated here, the proposal is written with certainty and clarity, a contractor will be placing most of the odds in his favor.

One further point to note is that once a contractor has decided on which plateau to bid the job, further cutting the bid price or staying at the bottom end of that plateau does not greatly enhance his chances of getting the job. In fact, he can speculate his bid price all the way down to close to nothing and still have no chance. The contractor's judgment of the competition and his knowledge of the customer are the best tools that can guide him in this endeavor. However, it's worth noting that if a contractor contracts, say, eight jobs per month and drops an average of $500 per job, he need not be an accountant to know the effect it will have on his bottom line. Being a low bidder is what our business is about, but being too low a bidder most of the time is an unnecessary burden.

A contractor should be attentive to this subtle principle, for after all it is not so subtle. Major corporations use it in their national marketing campaigns to lead the public to believe that they are offering lower prices when in reality they are using the higher end of their pricing structure.

WHAT ARE YOUR ODDS?

The odds are what a contractor makes them. Why should his bid be favored over another? A long list of answers may apply, but the most logical are lower price, reputation, and perseverance, with perseverance leading the pack.

Diligently following the development of a bid and providing the customer with the necessary estimating support from the moment a contractor accepts the invitation to its completion is a sure way to place the odds in his favor. The support referred to here mainly consists of the timely submission of a written quote, attending pre-construction conferences, and a genuine concern for the job.

A written proposal can be submitted well before the bid date. A contractor can, and should, submit its outline, short of the price, soon after he begins estimating the job. Aside from his standard terms and conditions, it should list the scope of work and exceptions, a task done with a standard form (see Chapter 3), giving his prospective customer ample time to digest the bulk of the bid. At bid date, the only thing left for the contractor to do should be to call in the price. Employing this method most likely will maintain a favored place for the contractor in his customer's preferred bidders list.

CHANGING THE BASIC PRICE STRUCTURE

In this chapter, the subject of fine-tuning the bidding price—for example, from one plateau to another—should not be confused with a fundamental change of the estimating pricing structure. Developing a specific pricing structure that keeps a contractor competitive and profitable is the result of a painstaking process. Once developed, the pricing structure is the contractor's foremost resource, and therefore it should be exempt from any tampering or speculation.

The bidding price should always be compiled from the estimated cost and the company library of markups and not from hearsay, guessing, or gut feeling numbers. Pursuing over-inflated bids, even when a contractor gets few jobs, ultimately will destroy his ability to compete in a realistic world. And when push comes to shove, a contractor will be out of the mainstream.

Chapter 8

THE CONTRACT

STANDARD CONTRACT FORMS

The multitude of standard contract forms used in the construction industry today makes it impractical, if not impossible, to list them all in one chapter. While the most common were originally accepted as fair and impartial, thanks to the imagination of the overprotective or unscrupulous customer, they have evolved into detrimental legal instruments from which we must protect ourselves.

What follows, therefore, is a collection of the most elaborate contract clauses a contractor is likely to encounter. These contract clauses are quoted unaltered as they appear in various executed contracts. Each abstract is identified by a letter, and key words or phrases that adversely affect the spirit of the clause are printed in bold. Any comment that follows a clause is intended to give the contractor a sense of how to avoid these and other similar legal traps. In the process he'll learn how to promptly recognize and spot these "sore spots" in his contract readings.

ABSTRACT "A"

Payments

- If the Owner does not pay the Contractor **for any reason**, including the Owner's insolvency, the Contractor has no obligation to pay the Subcontractor.

Comment: What if the contractor is negligent in carrying out the work? Should he be exonerated from all obligations?

Change Orders

- Only changes authorized in writing and agreed to in amount prior to performance by the Contractor will be paid. It is agreed that if such change involves an order, directive, act or omission of the Owner, or any other third party, then the Subcontractor shall be limited in recovery to the amount received by the Contractor from said third party, **less its costs, markup, expenses, and attorney fees**.

Comment: What we can gather here is that the general contractor wants to make his profit and more, but would not allow the sub to make a profit. In fact he's going to charge the sub all his expenses plus legal fees even if he only chooses to consult his lawyer for whatever reason.

Completion

- Work will be completed on or before 00/00/00. **Time is of the essence.**

Comment: When a contractor encounters *"Time is of the essence,"* bells should ring, for this phrase has a specific legal significance. The customer is alerting the contractor that the satisfactory completion and acceptance of the contract is contingent on the completion of the work on time. If the contract is not completed in time, then the contract has no existence or value.

Termination

- (a) Should there be any work stoppage or slowdown caused by a strike, picketing, boycott, or any other voluntary or involuntary cessation, delay or interference of work by the Subcontractor, or the Subcontractor's suppliers, or any of them for any reason, and which, **in the judgment of the Contractor** is causing or **is likely to cause** delay in the progress of the work to be done by the Subcontractor, the Contractor shall have the right to declare this Agreement in default in accordance with the procedures outlined in Sub-Paragraph (c) of this Paragraph 4."

- (b) All work under this subcontract shall conform to all applicable laws, rules, regulations and codes. Wherever the specifications or drawings conflict with such documents, **work shall be performed in compliance therewith at no extra cost.** Subcontractor further agrees that [it] shall comply with the Occupational Safety and Health Act of 1970 as amended and all rules, regulations, standards and orders issued there under. If any act or failure by the Subcontractor results in the assessment of a civil penalty against the Contractor, Subcontractor shall reimburse Contractor the amount paid by Contractor on demand by Contractor. **If, in the opinion of Contractor,** Subcontractor is in violation of any standard, rules, regulations, or code requirements of said OSHA Act, Contractor shall have the right to declare this Agreement in default in accordance with the procedures outlined in Sub-paragraph (c) of this Paragraph 4."

- (c) In the event that Contractor **desires** to declare this Agreement in default for any reasons stated in Sub-Paragraph (a) or (b) above, or in the event that Subcontractor threatens to or fails in any manner to properly or timely perform any obligation on its part to be performed under this Agreement, Contractor shall have the right to declare this Agreement in default after giving the subcontractor twenty-four hours notice of the intention of the Contractor to claim such default. . . .

Comment: These clauses, which unfortunately are becoming very common, should warn a contractor to read carefully, especially the bold typeface writing, and then decide how far he is willing to commit himself.

Subcontractor Representation

- The Subcontractor shall have a representative competent in the field of work covered under this Agreement on the job site at all times work under this Agreement is being performed. Said representative shall have full authority to act on Subcontractor's behalf and **shall not be changed without written consent of the Contractor**.

- Subcontractor agrees to indemnify and hold Contractor harmless from any and all liability, claims and damages, including attorney fees, which any party might seek or claim against the Contractor, or which it may incur, and which directly or indirectly, actively or passively, was caused by Subcontractor or one in its employ, contract, or for whom the Subcontractor might be responsible. Subcontractor's liability and indemnification shall be valid whether or not the Contractor was directly or indirectly responsible for the damage or was otherwise liable to the third party.

ABSTRACT "B"

Contractor-Subcontractor Relationship

- Contractor shall have the same rights and privileges as against the Subcontractor herein as the Owner in the General Contract has against the Contractor.
- **Subcontractor acknowledges that he has read the General Contract** and all drawings and specifications, and is familiar therewith and agrees to comply with and perform all provisions thereof applicable to the Subcontractor.
- The Subcontractor agrees to indemnify and hold harmless the Contractor and the Owner from and against any and all suits, claims, actions, losses, costs, penalties, and damages of whatever damage kind or nature, **including attorney's fees arising** out of, in connection with, or incident to the Subcontractor's performance of the subcontract work.

Comment: While this Hold Harmless clause appears to be milder than the one in Abstract "A" it is very bold when considering it was part of a $1500 contract—a steep liability to assume when the cost of attorney's fees for any claim can well be ten times the contractor's anticipated profit.

As subcontractor, a contractor must keep these clauses in proportion to the contract amount and his anticipated profit. A contractor should think! How can he prevent anyone from suing the owner for a defective device the contractor did not manufacture? How can he then indemnify the owner and the general contractor for something that is specified by the owner and manufactured by someone over whom he has no control? You make the final call.

- In the event that any work fails to conform to the requirements of the contract, the same shall be corrected by the Subcontractor immediately upon its discovery; further, in the event that defects due to faulty materials or workmanship appear within one year from **useful occupancy,** the

Subcontractor shall at his own expense correct the same and this applies to work done by his Subcontractor(s), his employees and those working by, through, or under him.

- The work covered by this contract shall be commenced within three (3) days after written notice to the Subcontractor by the Contractor, and completed within 60 days from date of commencement: **time being of the essence** of this contract. Loss of time on account of materials or work being condemned will not be considered as a cause for an extension of the contract time. In the event the Subcontractor fails to complete on time, he shall pay the Contractor by way of liquidating damages the sum of $300.00 per day for each and every day the work remains incomplete unless the cause for delay is caused by strikes not caused by Subcontractor, fire, governmental regulations, or acts of God, in which case the time for completion shall be extended for the actual time of such delay. Said liquidated damages shall be deducted as such from balance due the Subcontractor. Should damages exceed the sum due, or to become due, the Subcontractor, then in that event, shall be liable unto the Contractor for such difference.

Comment: Again, $300 per day on a $1500 contract?

- Should the Subcontractor fail to do or perform the work required hereunder, thereby in the **opinion** of the Contractor causing or threatening to cause delay in the general progress of the work, Contractor shall have the right to declare this Contract to be breached by the Subcontractor and cancel this Contract by notice to him in writing and to re-negotiate and re-execute contract(s) for the completion of the work required to be done under their Contract with such persons, firms, or corporations as shall in the **opinion** of the Contractor be necessary. In the event Contractor cancels the Contract, as hereinabove provided, it shall pay to the Subcontractor only for work completed to the date of the cancellation and the Subcontractor shall not be entitled to profits of any kind or character, anticipated or otherwise, or compensation for materials and labor unfurnished. Further, all losses, damages, and expenses, including attorney's fees in the prosecution or defense of any action or suit incurred by or resulting to Contractor on the above account shall be borne by and charged against Subcontractor, the Contractor may recover on said bond, and both Subcontractor and/or his surety agree to pay Contractor immediately such losses, damages, expenses, and attorney's fees.

Comment: In this clause a contractor is at the mercy of the customer's opinion, and not of accepted business practice.

- The Contractor shall not be responsible or accountable for any losses or damages that shall or may happen to the Subcontractor's work, materials tools, or equipment employed during construction until **all work is completed and accepted by Owner.**

Comment: What does it mean "until completed"? Would he pay then for damages?

- The Subcontractor shall put himself in communication with other subcontractors whose work may affect his, so as to promote harmony of work. In the event Subcontractor, its agents, employees, or subcontractors commits or allows to be committed any act or acts, or does or allows to be done any thing(s) which tends to create or creates disharmony, a work slowdown, work stoppage, or strike, then and in that event **the Contractor shall have the absolute right to immediately cancel this Contrac**t and complete same in accordance with breach of contract procedure as set forth in Section 10.

- All labor employed by the Subcontractor upon the job-site will be subject to the Contractor's approval and in the event the Contractor finds fault with any employee **for any reason whatsoever**, said employee shall immediately be removed from the job-site and replaced.

Comments: For any reason "whatsoever." Can a contractor afford the customer chasing his workers away?

- The work shall be done subject to the final approval of Contractor and Architect, if any, and their decision as to the performance of the work in accordance with the drawings and specifications and the true construction and meaning of same shall be final.

INDEPENDENT CLAUSES COLLECTED FROM VARIOUS CONTRACTS:

Payment Sample 1

- The Contractor shall submit to the Owner's Representative, before the Owner's Representative shall be required to make any payments, an application for each payment and receipts or vouchers showing his payments for materials and labor and a schedule of the various parts, including quantities, and supported by such evidence as to its correctness as the Owner's Representative may require; and the Owner shall at all times be entitled to retain 10% of all moneys due and owing to the Contractor as part security for the faithful performance of this agreement, said 10% so withheld shall not be paid to the Contractor until 30 days after final acceptance of the work or materials **shall have been made by the owner.**

Comment: What if the owner never accepts?

Payment Sample 2

- Project is to be invoiced at completion of project. A notarized Waiver of Lien must be submitted with Subcontractor's invoice. Approved Change Orders are to be invoiced separate from the contract amount and a separate lien waiver must be submitted. **No invoice will be processed for payment until a lien waiver** and certificate of insurance is received. Two payments will be issued: first for 90% of the contract amount and second for retention and approved Change Orders. Payment for properly submitted invoices will be made within five (5) days after Contractor's receipt of payment from Owner.

Comment: If in order to get paid a contractor must release his rights to lien, then why bother with lien laws in the first place?

WORKING CONDITIONS:

- Working hours are 12AM to 12PM or as approved by the superintendent.

Comments: Under this clause a contractor will have a tough time collecting overtime or any other premium time.

POSSESSION PRIOR TO COMPLETION

- Whenever it may be useful or necessary for the Contractor to do so, **the Contractor shall be permitted to occupy and/or use any portion of the work** which has been either partially or fully completed by the Subcontractor before final inspection and acceptance thereof by the Owner, but such use and/or occupation shall not relieve the Subcontractor of his guarantee of said work and materials nor of his obligation to make good at his own expense any defect in materials and/or workmanship which may occur or develop prior to Contractor's release from responsibility to the Owner. Provided, however, the Subcontractor shall not be responsible for the maintenance or such portion of the work as may be used and/or occupied by the Contractor, nor for any damage thereto that is due to or caused by the negligence of the Contractor during such period of use or occupancy.

Comment: The ceiling electrical rough is completed before the ceiling grid is installed and the owner ships all wall cabinet units. He invokes this clause, storing the cabinet units in the middle of the floor. When the grid is finally installed the contractor is asked to install the lighting fixture. The difference now is that the contractor cannot use rolling scaffolds and his crew has to work on 12-foot step ladders. If a contractor puts in a claim for it, the customer will reject it, for he says the contractor is not entitled to extra compensation. What do you think? Can the contractor prevent this from happening to himself? Yes! The contractor should be aware of the existence of this clause and then have it struck out before he accepts the job.

Owner's Furnished Materials

- Subcontractor shall be responsible for unloading, storing, maintaining, and inventorying any Owner supplied material and equipment pertinent to his trade. Written notice shall be given to the Contractor within 24 hours of any shortages or damage of Owner supplied material and equipment. **Failure to notify Contractor of shortage or damage within 24 hours** constitutes acceptance by the Subcontractor and the Subcontractor will assume responsibility for shortage and damage.

Comment: The only problem with this clause is that most owner-supplied materials arrive well in advance of schedule and the super or the owner doesn't want any boxes open until installation time. Have the clause struck out or the deadline extended to installation time.

Extra Work Clauses:

- If the Owner requests changes by altering, adding or deducting from the work to be performed under this contract, the Contractor shall give the Owner's Representative a written proposal outlining work to be performed, including cost of the labor, materials and equipment, which shall be approved by the Owner's Representative, in writing, before the Contractor proceeds to execute the work, and in any event the Contractor shall not be entitled to any claims of extras unless approved by Owner's Representative in writing. **Extra charges shall not be claimed by the Contractor due to weather conditions or job scheduling.**

Comments: The last sentence takes away the spirit of the clause, which is fair compensation for extra charges. What if the general contractor chooses to schedule work in such a way that it will impair the contractor's productivity far below his estimate? Should he be compensated?

POINTERS AND OBJECTIVES

Remember these are suggestions based on the legal doctrine that most courts of law accept. For specific and legal advice on these points of law a contractor should always consult his lawyer.

Construction Drawings- When a contractor receives a set of drawings, there is an implied guarantee by the owner that the drawings are correct and suitable to carry out the work. This is why a contractor cannot hold liable the architect, the engineer, or other professionals, except the owner. The only time a contractor can hold the architect or other professionals liable is when the damage involves injuries to others or to property. All other losses, such as money loss, contractors' loss of productivity, etc., are not binding to their performance.

Architects- Most architects are supervisory architects with limited control. Before a contractor tries to sue an architect, it is wise to know that in most cases recourse to professionals is a limited right and the "economic loss rule" applies. The owner is liable for defective design.

Judges- Clarity of contract is the contractor's best prevention. Most judges are not construction people. And construction disputes can be complex.

Beneficiary to the Contract- A contractor should always find out if there is any beneficiary to the contract he signs. There can be some nasty surprises if the beneficiary is not the person a contractor thought he was dealing with.

Right to stop work- A contractor should always negotiate this clause into his contract: "If not paid within 7 days the subcontractor has the right to stop work." Then if the owner breaks his promise a contractor can stop work.

Requisition- Contract breakdown for requisition (see Chapter 12) by virtue of clarity supports breach of contract. For example, the amount due is better defined. It gives less excuse for the owner not to pay and eliminates ambiguity, which is the cause of disputes.

Arbitration rather than court- When reviewing the contract a contractor should look for this provision. If it is arbitration, is it general arbitration or just for a specific dispute? The difference between court and arbitration is that the court will allow a contractor to call anyone he wants or that has anything to do with the case or dispute, while in arbitration the contract will say how to dispute. If the contract calls for arbitration, the contractor as subcontractor will be subject to arbitration. Arbitration is final; court can be appealed.

Termination and specifics- For example, giving notice. If a contractor wants to preserve his rights, follow the contract procedures to the letter.

Other contracts- The contractor should always check what other contracts or documents are part of or are incorporated in his contract.

Final Payment- The contractor should always negotiate how many punch lists he will be subjected to and whose punch list is final in order to release his payment.

Signatures- Photocopied or faxed signatures are okay until challenged.

CONCLUSION

Before a contractor undertakes a contractual obligation, he must have a clear perspective on these and other elements that entrap the unwary contractor, for the danger of running into crooked dealings will always be there.

A common myth among unwary contractors is their belief in standard contract forms. They are intimidated when exposed to long double-sided pages of small print, leading them to accept any such form as "standard" (whatever standard means) and so long as they are standard, their reasoning goes, it's okay to sign. And so they do. Nowadays, word processors, laser printers, and fancy-bordered color paper can produce the most standard-looking contracts money can buy.

A contractor, to free himself from this form of intimidation, should dissect the contract by first scanning through and highlighting the titles of those clauses that concern him most (If need be, he should review Chapter 3 and the earlier comments here), then he should review their content and underline or annotate his objections to each clause. Upon completion, he should analyze his objections and decide whether he wishes to negotiate the contract or simply walk away (see Chapter 9, "Negotiating the Contract").

A contractor should never underestimate the responsibility he assumes and the risk he and his entire staff take when signing a contract. Every one's future or job can be on the line. A clause that may appear trivial or misplaced has no meaning until invoked. A contractor should communicate with his super, estimator, secretary or anyone else involved in the outcome of the contract. A simple thought or suggestion from those he least expects can change things for the better. If there are doubts, a contractor should have such clauses crossed out if possible, and, if not, then he should take legal advice before he signs such an agreement.

All that has been covered so far in this chapter is not meant to paralyze the contractor nor prevent him from signing contracts and taking jobs. On the contrary, it's meant to encourage him to take jobs that he would otherwise walk away from for fear of not understanding the contract. The intention here is to give the contractor the ammunition to be able to negotiate his way into sound and fair contracts and avoid those sure losers.

In our industry the number of good customers far outweighs the number of bad ones. It just happens that the bad ones are easier to reach for they are always on the lookout for new contractors and appear to be greater in number. A contractor's challenge, therefore, is to find those good ones who are tightly held captive by his competition.

Chapter 9

NEGOTIATIONS

MEETING THE CUSTOMER

This is the day a contractor has been longing for, the day of negotiating and signing the job. Negotiating the job, however, goes beyond the legal ramifications of a contract. It involves translating the parties' offer and acceptance into a fair and equitable agreement.

In life, we constantly negotiate for what we want. *"I'll give you this if you give me that."* But wants, most often, come with strings attached. We must exchange things we already have for those things we want. Those that know how to ask and exchange usually get what they're after. In contracting, a contractor must sharpen his natural skills and, beforehand, know what he wants and its relative value (what he is willing to pay or give up for it).

Before we enumerate the most common items a contractor should bargain for when negotiating a contract, let's analyze the reason why he is meeting with the customer. The most obvious reason is that both of them think he can deal with the other. In other words, there is a meeting of the minds on two of the three basics of contract: solicitation, offer, and acceptance (see Chapter 3).

The solicitation, invitation for bid, was accepted by the contractor. The offer, the contractor's bid price, is (almost) acceptable to the customer. And the acceptance of the terms and conditions under which the contract is to be carried out is open for negotiation; that is the basis of the meeting. In order to close the deal, each is willing to give and take a few points.

Going to such a meeting unprepared is counterproductive, to say the least. To no one's surprise, what the customer mostly wants is a lower price and better legal protection. To the surprise of many, the customer, in return, is willing to compromise on the scope of work as well as on the terms and conditions. This opportunity comes once in the life of the contract, and it is before the contractor signs it. Any future modification can only be made through change orders, which usually carry heavier price tags.

An astute contractor should always be ready to negotiate most contracts without a rehearsal. Mastering this skill is quite simple. As a negotiator, a contractor should develop two mental lists: one for the general conditions and one for the scope of work (guidelines for these lists are detailed later on in this chapter). Each list should deal with items that are most likely to be accepted by the customer and that the contractor is willing to trade in for a fair compensation. A contractor should assign to each item a dollar value and use it as a bargaining chip to either get what he wants or to offset any price cutting. Each item's value will determine its negotiating priority.

However, before we proceed with this bargaining concept, we need to clarify one important point. The quoted price must include every item the contractor is putting on the table or he will be giving away moneys that were not allocated in his bid price.

With a little practice a contractor will be in control of all his bargaining chips and always able to get something in return for every dollar he shaves off his bid price.

General conditions:

These items are listed here because they generally favor the writer of the contract document—the customer—whose standard contract form most often is written to protect against an array of circumstances; unless a contractor points out its lopsidedness, it will stay as written. Staying within the spirit of negotiation, the contractor should negotiate toward his corner, or at least to the middle, those items he feels will benefit him most. A contractor will be surprised how many will fall into his corner just for the asking.

- **Method of payment**
  Whatever payment schedule a contractor works out is fine as long he doesn't burden his cash flow and as long as the balance due upon physical completion does not exceed 10% of the contract price.

- **Down Payment**
  A contractor should use his judgment and should not be too bashful to ask, especially if he thinks the job warrants such a payment.

- **Retainer reduction**
  A contractor should be aware—most contracts allow the original 10% retainer to be reduced to 5% or less after the job is 50% complete. A contractor should have that stipulated in his contract.

- **Substitutions**
  There is no better time to have specified material or equipment substituted than at the negotiating table. Because this can be a high-priced ticket item, a contractor should know his facts and figures beforehand.

- **Special Equipment**
  Through this item a well thought-out presentation can pay large dividends.

- **Limit the Change Order Amount**
  There are times when large change orders can adversely affect a job. For example, a contractor signed a $40,000 contract to wire a commercial shell that he now wishes he had not signed. The customer, in accordance with the terms of the contract, issues a change order to wire a tenant space on a time and material basis which is worth $90,000, but, for whatever reason, the contractor doesn't want to do it. The customer can hold him liable for breach of contract, thus jeopardizing the base contract.

  To prevent this situation, especially in projects where large change orders are probable, a contractor should evaluate all possibilities and then, if necessary, have a clause added to the contract requiring change order amounts not to exceed, say, 30% of the contract amount and giving the contractor the right to refuse if they do.

- **No back charges without notice**
  Customers back charging contractors at the end of the job is a common practice that is difficult to fight, especially when the prospect of collecting the final payment is overshadowed by legal fees that can be larger than the amount due. A fair and easy safeguard is to add a clause that reads something like this: *"The customer shall notify the contractor of any*

impending Back Charges at the time they occur. Back Charges shall be treated as Change Orders."

- **Completion date**

  A contractor should bear in mind that the electrician, in most jobs, is the first one in and the last one out. His completion is mostly dependent on others. If a contractor can, he should stay away from fixed dates. They usually breed liquidating damages.

- **Premium time**

  This is a very costly item if not clearly defined. A contractor should be careful in his negotiation. In many cases, it can easily get out of control.

- **Surplus Material**

  In alteration work, it's always good to know who owns surplus material. A contractor should state it clearly, because it's worthless until a salvage company makes an offer to the customer.

Scope of Work

A contractor should know the cost, when applicable, of each of the following items before he begins negotiations.

- Survey work: Benchmarks and coordinates
- Concrete cutting and core boring
- Trenching, back-filling, and compacting
- Cutting and patching
- Garbage removal
- Fixture supports
- Lamp installation
- Fireproofing
- Hoisting and warehousing
- Maintenance of temporary light and power

Negotiating fair contracts is an exciting challenge. A good negotiator is always respected and ultimately makes good, lasting friends, especially when he delivers on his promises.

Chapter 10

PROTECTING THE JOB

SETTING THE STAGE

Because unscrupulous contracting is as detrimental to our business as cancer is to society, we have to learn how to prevent it, and, if necessary, how to deal with it. It's important to note that while we're most concerned with unscrupulous contracting, disputes can arise in most relationships.

To protect his interests in every job he does, it's better for a contractor to have a standard policy of recording the most prominent events of a business relationship in chronological order from the start than to try to trace events back when the contractual relationship has deteriorated into a dispute and everyone has gone home at the end of the job.

The ultimate answer to a dispute is to dispute it in a court of law. However, the most practical and often the most profitable way is to prevent the dispute before it takes place—that is what this chapter is all about.

It's not the intention of this work, however, to precipitate the contractor neither into an array of legal maneuvers nor into starting a war with his customers. When a contractor adheres to basic policies, he will select and preserve items of contractual importance for his files. And should the need arise, he will have a set of legal documents capable of delivering the one ounce of cure that settles disputes out of court.

DOCUMENTATION

To document the job's most pertinent information, drawings aside, a contractor needs a set of files where information is compiled and maintained for future use. Typically this set consists of:

- Job Folder
- Diary
- Intelligence File

Job folder

An efficient job folder for most contracts is a six-section legal folder. The sections can be labeled as follows:

- Estimate/ Contract / Permits / Correspondence
- Contract Breakdown / Billing
- Pending Change Orders
- Approved Change Orders
- Submittal Pending (Letter of transmittal only. Shop Drawings and the like are filed as drawings.)
- Submittal Approved (Letter of Transmittal Only. Shop Drawings and the like are filed as drawings.)

Diary

The diary, when accurately kept, is the backbone of the contractor's stage-setting procedure. On large jobs with a field office, the diary is kept by the job super or foreman. However, in smaller jobs, where the labor force attendance is periodic, the safest procedure is to keep the diary in the main office and have the lead-men report in at the end of each working day. A contractor should remember that field people have a greater tendency to procrastinate on paperwork and tomorrow, not to mention the day or week after, their recollection of events is not as sharp as it is today. With every passing hour, their recollection can dwindle down to nothing.

The diary should be standardized to include basic topics, giving uniform guidelines to the person making the report and minimizing oversights. A good diary, aside from its obvious advantages for recordkeeping, enables the contractor to follow the job progress and recount each event with accuracy. It can be a very influential record in a court of law.

The most economical and credible diaries are those maintained in inexpensive spiral-bound notebooks. A contractor should maintain one for each job. Each page should bear the same format:

- Date and day of the week
- Weather, including the day's temperature
- Workers' names and hours worked
- Items worked on
- Change orders and directives issued or received, if any
- Material received
- Inspections
- Visitors' names
- Rental equipment received or returned

Intelligence File

Most newcomers to this concept may frown at the thought of setting up an intelligence file, especially on business associates. The fact is, however, that there are many who wish they had done so long before they got hurt.

This legal size manila folder should contain confidential information about your customer and it should be kept off limits to others. The most common information compiled in this file is:

- **Postal stamped envelopes-** where a discrepancy exists, or is suspected, between the date received and the date claimed in the contents of the envelope.
- **The customer's first transmittal-** where a contractor is invited to bid the job and it lists the drawings he was sent. This is strong evidence when the contractor is asked to perform work shown on drawings he didn't know existed.
- **Photocopy of the customer's check-** Suing deadbeat for money is an inevitable fact of life that sooner or later a contractor will face. Having the customer's banking record handy will greatly facilitate that process. Getting a judgment, at times, is easier than getting a bank account record. Therefore, a contractor should keep an eye on all the checks he receives; they can reveal crucial information about the customer and what he is doing. For example, bank and payer name changes can in some cases

reveal a deceptive transition to an entity that has no legal ties with the contractor's contract. Checks that he receives from other parties on behalf of the customer can have a special legal significance. In essence, a contractor should preserve photocopies of all the different bank drafts he receives because once he deposits them their valuable information is no longer in his possession.

- **Credit reports, if any**
- **Customer personal data-** Home address, phone, etc.
- **The customer's super-** Home address and phone number, or at least the super's vehicle license plate number and model. The most honorable job superintendent at times gets fired in the middle of the project, leaving behind an array of commitments the customer most often will not honor. When this occurs, this superintendent would welcome the opportunity to defend himself and his commitments out of moral obligation, if only he were asked. However, industry practice and contractual obligations with his former employer prevent him from freely coming forth on the contractor's behalf. Obtaining his address and phone number, should the occasion arise, allows the contractor to get in touch with him promptly, rather than having to track him down through costly and time-consuming legal procedures.
- **Auto and truck license plate numbers**
- **S S # and driver's license numbers**
- **Occupational and construction license numbers**
- **Other subcontractors' data-** Recording other subcontractors' names and phone numbers in this file can help a contractor communicate with those who may experience the same problem he does. Collectively, they may be able to help each other.

Many of these items may seem to be hard to get, but as a contractor sharpens his skills he'll become more vigilant and begin to notice that most of this information, at one time or another, flows right through his office. For example, permit applications, notices of commencement, occupational licenses, and the like, are abundant sources that a contractor can easily access during construction.

FILING NOTICES

Each locality has its own set of rules when it comes to Mechanic's Liens law. If a contractor complies with their rules, and files the appropriate notices in a timely fashion, he will retain valuable rights that will help him protect his investment and collect his money. It is advantageous to learn the local rules and follow their procedures to the letter. In most areas, these procedures can be obtained from the local building and zoning department.

The procedure usually involves the filing of a Notice to Owner and, if a contractor doesn't get paid for his work, of a subsequent Mechanic's Lien within a specified number of days of the commencement and completion date, respectively. In many counties there are specialized private companies performing this service for a nominal fee. This is the best option for most contractors. See Chapter 16, "Collecting Your Money."

However, no matter who does the legwork, it is imperative that a contractor file all notices on time. The "Notice To Owner" procedure, for those who feel they are offending their customers, is not offensive to those customers that intend to pay; in fact, it's accepted for what it is—a contractor's legal right to protect his interests.

OTHER FORMS OF PREVENTION

Unscrupulous customers, like predators, have a craving for wounded prey. Therefore, anyone who is or appears to be wounded becomes fair game to these predators. That is why the ploy of claiming poverty, used by many contractors, most often backfires and hurts them.

The moral here is that a contractor should not resort to unnecessary claims when demanding what is rightfully his. In fact, under no circumstances should he ever invoke extraordinary claims. When asking for moneys due him there are two fundamental things a contractor should never say or do: (1) Never explain why he needs the money—the contractor earned the money and it is due him is the best explanation; and (2) he should never say "I'll sue you" or any similar-sounding phrase. A contractor should avoid deadlines—the customer is well aware of the contractor's rights, and the contractor can do as he pleases anytime. Saying "I'll sue you" to some can only be music to their ears.

A contractor telling his customers his financial problems, in the long run, does more harm than good, even when it appears to have been received well. No matter how a contractor looks at it, when he claims hurt to his customers he's a wounded prey. If the customer is unscrupulous, the contractor just walked into his den and, if he's not, this may be the last job the contractor will do with him. Being wounded, or just playing the part, is a losing proposition either way.

Therefore, another good safeguard is to always be financially healthy, if only in the customer's mind's eye. If the customer knows the contractor has the financial strength to wait out the legal procedures, the contractor's chances of collecting are greater than if he claims poverty.

Chapter 11

CONTRACT MANAGEMENT

JOB COMPLIANCE

The management of a contract, if its paperwork does not interface with previous work, can be inefficient and complex. If from the acceptance of a sales lead to the signing of a contract we use sound managerial procedures, then all subsequent tasks such as contract management and payment requisitions are simplified for they are a continuation of those procedures. For example, if a contractor sets an estimating folder with data from his sales lead and compiles his bid data from his estimating folder, he then has the bases for an efficient job folder that is ready and easy to manage, similar to the one described in Chapter 10.

In an ever-demanding industry like ours, this method, besides using information that is otherwise wasted, allows the contractor to begin the contract administration promptly and mobilize the job smoothly. Any form of procrastination, at this stage of the job, only breeds unwanted and costly delays.

To help the contractor start, the following lists have been compiled as guidelines for the most common items of work. For specifics, a contractor should read his contract documents, paying special attention to deadlines and format requisites. For instance, some customers will accept payment requisitions and releases of lien only on specific standard forms:

THINGS TO DO

- **Permit Applications-** The contractor should comply with his local Building and Zoning requirements by filing all necessary applications for electrical permit(s) in a timely way.
- **Insurance Certificates-** The contractor should have his insurance agent send certificates of workmen's compensation and general liability to his customer. Make sure they comply with the specified limits and they are made out to the correct party. Also, check for other insurance requirements such as automobile and other special coverage.
- **Notice to Owner-** A contractor should protect his Mechanic's Lien rights. Whenever applicable, he should comply with his state requirements. Don't overlook this important procedure.
- **Utility companies-** A contractor should notify or apply to those utility companies that provide services for his contracted work, such as electric, telephone, cable, security, and the like.

SUBMITTALS

- **Shop drawings and samples-** A contractor should follow the chain of command and submit, within the specified number of days, the correct number of copies to the architect. Any material that is subject to the architect's approval should not be released until he obtains such approval. A contractor should pay close attention to the "Stamp of Approval." Some shop drawings are "Approved as Noted," which means their approval is conditional on the architect's notations. The contractor should transmit a copy to his vendor and release his material subject to his compliance to the architect's approvals and notes.
- **Substitutions-** Whenever a vendor wants to substitute a specified item and the contractor is willing to accept the substitution, then the contractor should make these submissions early on—preferably before he starts the job, for the processing time can be extensive.
- **Payment schedules-** When required, the contractor should submit his contract breakdown soon after he signs the contract (see Chapter 12).

SPECIAL DOCUMENTATION

Nowadays, due to various federal acts, similar jobs may be subject to different rules, especially when built in different localities. Whenever a contractor is involved with these jobs he should be sure to maintain and promptly file the necessary reports; for example, Weekly Payroll Reports, OSHA, Environmental, Affirmative Action Reports, and the like. The contractor's failure to comply can constitute a breach of contract.

CHANGE ORDERS

The report of an event that changes the contracted scope of work is called a change order. The contractor's managerial skills in detecting, writing, and reporting these changes can greatly affect the job profit.

To effectively validate changes, we need to apply certain guidelines similar to those accompanying the change order form shown in Chapter 4 and to those that follow in this chapter. When a contractor does so, his contract administration is effective and productive.

How many change orders

A contractor should write as many change orders as it takes to expedite their approval, their completion, and their collection.

Customers and designers typically are reluctant to accept change orders, especially those that add cost to the job. Therefore, in an attempt to prevent or minimize the cost of the much dreaded change order, special terms and conditions restricting markups, accounting methods, and the like are constantly updated, making contracting ever more complex.

Although we dread change orders, we must be realistic and let experience guide us in this inevitable event. And let's learn how to protect our interests by using methods that facilitate and secure payment on all extraordinary work, even if we have to resort to psychological approaches to minimize cost impact and to ease the process.

Writing a change order for each event and keeping unrelated subjects separate is fundamental to smart contracting. In addition to facilitating recordkeeping and expediting the process, it benefits the customer-contractor relationship that usually gets tarnished with each new change order.

Conversely, when several changes, well after their completion, are compiled into one change order with weak or non-existent documentation, the customer, as well the contractor loses perspective on the work, creating unnecessary and costly disputes.

Smart contracting understands the business the contractor is in and the position he holds as a subcontractor in relation to the general contractor, the owner, the architect and any others that have anything to say or do with the approval and collection of his money. Protect his interests by preventing any situation that may hinder a contractor from smoothly completing the job, collecting all his money, and preserving his good name.

The following is a list of items that most often fuel such disputes. The comments that follow each item illustrate how some preventive measures may help.

- **Pricing-** Most of us resent unexpected price increases. This resentment worsens with each dollar increase and with each passing day the contractor fails to report the change. Understanding and dealing with this psychological effect is smart contracting.

  If a contractor submits a lump-sum $10,000 change order at the tail end of a $30,000 contract, then most often, he will jeopardize collection by turning a good customer into an instant enemy.

  However, if a contractor promptly submits ten $999 change orders, performs the work only after each change order is approved, he will get paid cheerfully.

- **Clarity-** Writing a change order for each item of work allows the contractor to clearly describe the scope of work and other specifics that otherwise can be confusing and misleading. Clarity in describing changes should never be confused with brevity. At times, for clarity's sake, a contractor may have to write a lengthy explanation to give the reader a good perspective of the work involved and to justify the cost.

  For example, "Furnish and install 2-250 Amp circuit breakers enclosures in meter room." While brief, the description is a far cry from:

1 Scope of Work: Rework meter room as directed by the electrical inspector and as laid out by the architect to protect two existing feeders, each serving power panels A and B.

2 Remove 2 existing 2-½" RGC each with 4-#3/0 cu from the existing 400 A 3 Phase main disconnect.

3 Install 1 new 8" x 8" x 36" wireway inclusive of wire and splicing kits.

4 Install 2 new 400A frame enclosures each with 250 A 3 poles circuit breaker.

5 Rework and connect the 2-½" conduits (see item 2 above) into new disconnects.

The above case is an abstract of a true dispute where the second explanation caused the architect to change his position and approve the change order as re-submitted and as priced.

- **Paper Trail-** Individual change orders make it easy to track down specifics such as who ordered and approved the work or who worked on it and for how long.
- **Authorized Signature-** Usually authorized field personnel for the customer can only sign and approve change orders up to a certain amount. When a contractor submits lump-sum change orders, chances are they will exceed that amount, thus delaying the approval because the contractor will introduce other people who are generally remote from the field.
- **Collection-** In lump-sum change orders, if one item is in dispute then the entire change order is in dispute and held back until that one item is settled. On the other hand, changes individually listed on separate change orders can be settled each on its own merit, thus giving the contractor greater negotiating power.
- **Profit-** Individual smaller change orders offer better profit opportunities than large ones.
- **As-built drawings-** Changes to the scope of work, in most jobs, have to be recorded in "As-built drawings." Individual change orders facilitate that task. Each change can easily be tagged on such drawings with its referenced change order number.

In conclusion, while change orders are an integral part of contracting, they have a stigma which, when not properly handled, can become the main cause of disputes that affect and often destroy business relationships. When a change to the scope of work is ordered or surfaces, the contractor should adopt the individual change order method. Saving a few pennies worth of paper and a few extra minutes of writing definitely is not worth the consequences. With clarity, a contractor should write as many change orders as possible and he'll have smooth sailing.

BACK CHARGES

Back charges, for all intents and purposes, are changes to the contract and, as such, should be treated as change orders. The only difference is that the charge is a deduction instead of an addition.

Chapter 12

REQUISITIONS FOR PAYMENT

THE REQUISITION

As the contractor does his homework for change orders and contract basics, compiling requisitions for payment should be fun, for he will see and feel, first hand, the return on his investment. Preparing a requisition gives a sense of accomplishment similar to that of a farmer harvesting his fields. All that was carefully planted is paying off. How much it will yield, however, is dependent on how well the contractor worked his fields (see Chapter 13).

To prepare an effective requisition requires five basic elements:

- Contract breakdown or Schedule of Values
- Change orders schedule
- Percentage of work earned
- Statement of account
- Invoice

CONTRACT BREAKDOWN /SCHEDULE OF VALUES

A contract breakdown is the base used to tally up the percentage of work completed and the amount earned at any billing period. Because a contractor will need it to compute his requisitions, he should get his customer to approve it either at the signing of the contract or soon thereafter. An efficient way to generate a contract breakdown is at the estimating stage of the job. The examples that follow in this chapter are based on that sample estimate.

To acquaint the contractor with the essentials of a requisition, let's assume our sample job has been in progress a few months and pick it up at the second month. "Requisition No. 2" includes previous and present billing with pending and approved change orders—an entire cycle, if you will. The following comments pertain to the form shown in Figure 5:

Requisition period- When a contractor is submitting his first contract breakdown for approval it is important to note the word "Initial" in the Requisition Period line in the space provided after "( ) Other." The initial submittal should include only the basic contract breakdown.

Items of work- In this column the contractor may break down the contract amount into as many items as he wishes. For example, he may choose a short list such as rough, trim, and final or a detailed list such as service, feeders, branch circuits, lighting fixtures, low voltage and so on.

In preparing a requisition, the contractor's main concern should be how to minimize his cash outlay for the shortest time and how to facilitate collection. This process is best served when items of work such as special equipment are billed independently from installation or when larger items are divided into smaller increments.

In our example, we divided the switching gear into three sub-system amounts—11a, 11b, and 11c. In larger jobs you can further divide the system to allow, for instance, for final testing or any other subdivision the work may call for. Sub-system 11a, "switching gear on site," allows for partial payment requests as the switching gear is delivered to the site and prior to its installation. In our monthly requisition we billed "Last Period" at 60%, "This Period" at 20%, and "Total to Date" at 80%, the remaining 20% to be billed when the shipment is completed. Sub-systems 11b and 11c allow billing for the installation independently from the material on site.

Another example of how a divided item will help the contractor collect his money is System 12, "Branch Circuits." When this system is left combined—pipe and wire together—it can at times be difficult to obtain full acceptance for a partial requisition unless all wires are pulled and spliced. This drawback is mostly felt at requisition time when most of the conduit is installed and the customer, because the conduits are empty, substantially cuts the requisition, claiming the work done was at far less value than the amount requested.

As our example shows, anytime the contractor has items of work that for payment purposes are not clearly definable, instead of leaving it to the customer's discretion at requisition time, he should take control by dividing them into indisputable and easy-to-track increments.

The principle here is to facilitate the request for payment, which is basically achieved through clear accounting. The more a contractor singles out an item of work, the easier it is to track its progress and thus the amount he has earned.

Clarity in contract breakdown amounts can easily establish breach of contract, especially when an item of work is completed and not paid for—a plus in the contractor's corner.

Contract Amount- The contract amount for each item of work is equal to its prime cost plus a constant percentage of the difference between estimated prime cost less DJE and sales price. We compute this percentage as follows:

Difference = Sales price ($30,771) less prime cost ($21,007) less DJE ($1,759)   = $8,005
Percentage = $8,005 ÷ $21,007 x 100 = 38.106%

This percentage is the contractor's workhorse for converting estimated prime costs into contract amounts. Remember, sales price and base contract total are one and the same. In computing percentages, sales price is the ultimate amount shown on the contract and not necessarily the amount shown on the bid recap sheet.

Therefore, as a guideline to help us compute the contract breakdown amounts we simply add 38.106% to each prime cost item in the estimating recap sheet. When adding the "Contract Amount" column, you may have to round off some numbers to obtain the precise base contract total.

This percentage is also a helpful guideline in computing items of work that are subdivided, such as 11a, 11b, and 11c. For better control, a contractor may compute each contract amount by generating each prime cost first from his estimate recap sheet and then adding his percentage value to each subdivision.

Mobilization (DJE)- allows the contractor to request payment for direct job expenses such as permit fees, insurance and bonds, job site storage and the like in his first requisition or as he earns them. In our example, Mobilization equals Line 16 plus Line 24 of the bid recap sheet ($1675 +

$84 = $1759).

A well thought-out contract breakdown, when done skillfully, gives a contractor that much needed edge to offset some lopsided contract clauses such as 10% retainer until well after the job is 100% completed or the "Termination of Contract at Owner's Convenience." This is an opportunity a contractor cannot afford to overlook. The contractor should work his breakdown and diligently get it approved early.

John Doe Electric					Requisition No. 2		Page 1 of 3	
CONTRACT BREAKDOWN								

Job Name: Sample job 1, Warehouse — Job #02-1231

Requisition Period: (x) Monthly   ( ) Weekly   ( ) Other: Date: 02/28/20--

Prepared by: MS   Submitted to: Owner   Period Ending: 02/28/20--

	Items of Work	Base Contract Amount	Percentage and Value of Work Earned					
			Last Period		This Period		Total To Date	
			%	Amount	%	Amount	%	Amount
1	Mobilization/DJE	2,077.00	30	623.10	20	415.40	50	1,038.50
9	Service	4,579.00	20	915.80	60	2,747.40	80	3,663.20
10	Secondary feeder	5,296.00	30	1,588.80	50	2,648.00	80	4,236.80
11a	Switching gear on site	4,044.00	60	2,426.40	20	808.80	80	3,235.20
11b	Sw-gear back boxes installed	300.00			80	240.00	80	240.00
11c	Switching gear final	278.00						
12a	Branch circuit conduits	2,094.00	20	418.80	40	837.60	60	1,256.40
12b	Branch circuits wire	1,500.00			30	450.00	30	450.00
13a	Lighting fixtures on site	4,520.00						
13b	Lighting fixtures installed	960.00						
14	Wiring devices	598.00						
15	Equipment connections	649.00			20	129.80	20	129.80
20	Fire alarm	3,876.00						
	(This and other forms may be downloaded online at www.theestimatingroom.com)							
	Base Contract Total	30,771.00		5,972.90		8,277.00	46	14,249.40
	(From other side) Change Orders Total	985.00		125.00		465.00	60	590.00
	Revised Contract Amount	31,756.00		6,097.90		8,742.00	47	14,839.40

Note: Unit prices shown on this form are for billing purposes only. They don't represent the actual cost of the item and as such, they should not be used as base-cost for adding or deleting work from the contract.

Form EA-201-1-rev05 ©2005 The Estimating Room™ Inc.

Figure 5: Contract Breakdown

John Doe Electric							Requisition No. 2		Page 2 of 3		

CHANGE ORDERS SCHEDULE

Job Name: Sample job 1, Warehouse   Job #*02-1231*

Requisition Period: (x) Monthly ( ) Weekly ( ) Other: Date: 2/28/20__

Prepared by: MS   Submitted to: Owner   Period Ending: 2/28/20__

Change Orders Status						Percentage and Value of Work Earned					
Contractor's Data			GC	Add/Deduct		Last Period		This Period		To Date	
#	Title	Date	Approval	Days	Amount	%	Amount	%	Amount	%	Amount
1	Add interior lights	1/8/__	1/12__	0.50	450.00			40	180.00	40	180.00
2	Wire irrigation pump	1/16/__	verbal	0.75	625.00	20	125.00	60	375.00	80	500.00
3	Relocate power devices	1/22/__	pending	0.50	(350.00)						
4	Delete 2 outlets (Credit)	1/23/__	1/23/__		-90.00			100	-90.00	100	-90.00
5											
6											
7											
8											
9											
10											
11											
12											
13											
14											
15											
16											
17											
18											
19											
20											
	Total (Post to Requisition Form)			1.25	985.00		125.00		465.00	60	590.00

Form EA-201-2-rev05 ©2005 The Estimating Room™ Inc.

Figure 6: Change Order Schedule

CHANGE ORDER SCHEDULE

The change order schedule shown in Figure 6 is perhaps the most valuable sheet in the requisition set. With each submission, it provides the customer with a complete update on pending and approved change orders, thus barring him from any subsequent claims of unawareness, especially when he's remote from the job site.

In conjunction with the standard change order form and its log annotation as shown in Chapter 3, the change order schedule shown here provides the contractor with a simplified and effective method of recordkeeping that can support most administrative procedures—whether manual or electronic. With this thought in mind, we shall explore its most prominent features:

Change Order Status- columns "Pending" and "Approved" facilitate the tracking of change order amounts and extensions of time. When a change order is first submitted, its extended time and dollar amount are logged in the pending column. Upon its approval, these two values are transferred to the approved column from which a contractor can bill percentages of work completed at each requisition period.

Pending Change Orders- cannot be billed, nor can work start until formally approved by the parties. "Formally" means that both the contractor and the customer agree and sign to all the changes the change order entails, including dollar amount, extension of time (when applicable), method of payment, and scope of work. Short of this, unless the contractor has established a different procedure, working on pending change orders will give rise to a breach of contract, not to mention financial exposure—an agonizing and risky situation in which many well-meaning contractors find themselves.

In dealing with change orders as we know them, special attention has to be paid to the word "Order." A change has to be ordered, and it's not official until the customer issues a documental order.

STATEMENT OF ACCOUNT

The requisition cover sheet is a recapitulation of the contract herein called Statement of Account. Staying in tune with the concept of adopting a requisition format of his own, the cover sheet presented in Figure 7, thanks to its "Requisitions Recap" section, is the contractor's best insurance against oversights and mistakes when doing accounting.

Requisition Recap- This section is the checks and balances of the system. For example, if by mistake an item of work has been omitted or over billed, or a requisition underpaid, sooner of later it will show up in this section, making it impossible for it to balance.

When computing this section, it is a good practice to cross-check the calculations. For example, each line's "Last Period" plus "This Period" has to equal the "Total to Date" amount. If any one line doesn't add up, an error exists in the computation. The contractor should carefully cross-check each line and column until the requisition is balanced.

INVOICE

The next step is to write an invoice for the amount due "This Period." The description can be a simple line such as: *"Partial Billing, Requisition No. 2. Amount due this requisition $7,884.00"* If there is any balance due, it should not be added to this invoice. If need be, the contractor should just attach copies of those previous invoices. He should treat any invoice as the single statement on a specific portion of moneys earned.

The contractor should keep his accounts receivable straight; he should not bill more than once for the same work. The amount he can invoice with each requisition is the Amount Due shown in the "This Period" column only and not in the "Total To Date" or "Last Period."

The amount due shown in the "Last Period" column can be the result of any unpaid balance on any previous requisition. The amount due shown in the "Total to Date" column is the result of any balance due plus the present requisition.

John Doe Electric Co.
Licensed Electrical Contractor E-12345

123 Main St, Our City, USA, 12345, Fax (123) 456-7890

Requisition No. 2 Page 3 of 3

Date: 2/28/20__

Period Ending: 2/28/20__

(123) 555-6400

Customer	Job
Contact: John Smith	Job Name: Sample job 1, Warehouse
Name: XYZ Development	Contract No. 02-1231 Dated: 10/15/__
Add: 3456 Washington Street	Add:
City/St/Zip: Any City USA 12345	
Phones: Hm: Wk:	Owner: XYZ Development
Beeper: Fax:	Job Phone:

Documentation

Attached (X) Number of sheets including this sheet: 3   Not applicable ( )   Will Follow No Later Than / /

Contract Recap

Basic Contract Amount	$	30,771.00	Prepared by: MS
Approved Change Orders	$	985.00	Approved by: LS
Back Charges	$		Date Faxed: 2/28/____
Other (+/-)	$		Invoice No: 00234
Revised Contract Amount	$	31,756.00	Payment Due Date: 3/10/____

Requisitions Recap

	Last Period	This Period	Total To Date
Amount Earned	$ 6,097.90	$ 8,742.00	$ 14,839.90
Less 10% Retainer	$ 609.79	$ 874.20	$ 1,483.99
Net Amount Earned	$ 5,488.11	$ 7,867.80	$ 13,355.91
Less Payments Received	$ 5,000.00	$	$ 5,000.00
Other	$	$	$
Amount Due	$ 488.11	$ 7,867.80	$ 8,355.91

Submitted by: _____

Signature: _____Title_____

Form EA-220-Rev05 ©1994 The Estimating Room™ Inc.

Figure 7: Statement of Account

Standard Requisition Forms- In preparing requisitions, an important thing to take into consideration is the contractor's familiarity with the format he's using. The more a system is used, the more efficient people using it become, thus minimizing errors. This is why many customers impose a standardized form. While this makes the customers' lives easier, it leaves the contractor with different paperwork for each customer. This increases his workload and most often reduces his recordkeeping to a shambles.

To attain a level of proficiency in preparing and tracking requisitions, a contractor should standardize his internal paperwork, just as the customer does. Before he goes filling in any of the customers' forms, he should make it a policy to adopt his own standard format first on which to compile contract breakdowns, change orders, status reports, and requisitions. The contractor should use his format as the official paperwork from which he can check and submit requisitions, or, when requested, generate data that can be easily transferred onto the customers' forms.

Chapter 13

MOBILIZING THE JOB

IMPLEMENTATION

Mobilization as defined in the previous chapter is just another item for billing Direct Job Expenses. Mobilization as defined in this chapter includes elements affecting the performance of the job directly. How a contractor sets up the latter is vital to the final objective—making a profit on every job.

The actual mobilization of a job begins with the implementation of the administrative work (see Chapter 11, "Contract Management"), and it is followed by the physical mobilization of the job itself—the subject of this chapter. Procrastination and lack of direction at any stage of the work can be detrimental to the job's bottom line.

Before we begin with this study we need to emphasize that in our business "the job" is the main direct source of profit and because of this a contractor must always remain attentive to its needs. Nothing should take precedence over this. If you take two different contractors with opposing points of view on this subject and give them the same job to do, one will lose money and one will make money.

This chapter—in fact this entire section—deals with the elements of success. The cornerstone of contracting is the implementation of these elements.

In this Chapter we will cover:

- Job Layout
- Labor
- Material
- Tools

Job Layout

Most profits are made on the drawing board. Here is where the contractor's electrician's skills and knowledge of the applicable codes will pay many times over. The drawing board is the place where the contractor makes final adjustments to the scope of work and decides who should carry out the work and what method is the most productive.

Another advantage of the drawing board is that it allows the contractor to discover and correct mistakes and oversights before construction starts. When these errors are made by others we are usually entitled to compensation. When we are at fault, however, the repercussions can be costly back charges and delays.

A simple way of preventing most of these problems before we start a job is to scrutinize and review, not only the contract documents, but our estimate as well. A contractor should lay them out and tear them apart, if need be, so as to optimize anticipated profits and prevent any difficulties, so that in the end, when he delivers on the contract, all will balance out into a profitable result.

LIMITATION OF CONTRACTED ITEMS

The first and most essential part of laying out a job is to clearly define the contracted work so that all revisions and interpretations, including the latest agreements, are reflected in an easy to follow format (See Figure 8, below).

The objective here is: (1) to officially communicate to the field the latest revision in a concise format, rather than revealing copies of the contract or other confidential documents, (2) to prevent costly mistakes by pointing out special materials and contracted item pay-lines, (3) to prevent duplication, thus reducing the number of productive man-hours spent on research work, and (4) to increase productivity.

In adopting this method, besides closing an expensive communication gap that often exists between administration and production, the contractor will supply his lead-man with a worthwhile tool to help him expedite work decisively and effectively.

The sample form that follows is presented for its flexible style. In the "Work Description" column, a contractor may follow a format that synchronizes with a standard list, as shown, or freely list only those items that he's most concerned with.

The form can be used for the entire job, just part of it, a single change order or work order, or just daywork. Regardless of how the contractor uses it, when properly used, it will always deliver clear guidelines for a productive field directive.

LIMITATION OF CONTRACTED ITEMS OF WORK

Project: Sample job 1, Warehouse		Job No: 02-1231
Prepared by: MS	Date:	Sheet # 1 of 1
(x) Full Contract	( ) Partial: Syst. # _____ Title _____	C.O. #_____

Code: R = Remove F = Furnish I = Install C = Connect

System No.	Work Description — When applicable, note the type and size of material	Electrical Contractor				Others			
		R	F	I	C	R	F	I	C
9	Service: Trench and back-fill for lateral 36" deep						X		
10	Secondary **Feeders:** Substitute EMT conduit with PVC schedule 40. Use straps and back-straps every 36". Add #6 green wire		X	X	X	X			
12	Branch Circuits: EMT conduit Use compression fittings in warehouse and die cast setscrew in office area. Pull #12 a green wire in all conduits. Mud-rings 5/8" deep		X	X					
13	Lighting Fixture and Lamps		X				X		
14	Devices: Toggle switches and convenient outlets 20A 125V Plates: in office area ivory and in warehouse area brown.		X	X					

15	Equipment Connections: Cords and caps for plug-in
	                      Hard-wired connections
	                      Motor starters and disconnects
			X	X	X		X
							X
								X
								X	

17	Telephone: Complete system including electrical permit					X	X	X	X

20	Fire Alarm: Empty conduits and boxes
	              Equipment, wiring, and testing
	              Manuals, certifications, training
			X	X			X
							X
								X	X

(This form and others maybe downloaded online at www.theestimatingroom.com)

Form EF-501-rev05 ©2005 The Estimating Room™ Inc.

Figure 8: Limitation of Contracted Items

REDESIGNING THE WORK

Along with the contractor's field directive, a sketch or detailed drawing on how the work should be done will pay off handsomely. He should bear in mind that all that was designed was done by others remote from the job site and for a building that is not yet built. Here, before the contractor spends his first dollar, lies an excellent opportunity for him to rearrange an array of randomly drawn lines and differently shaped symbols, referred to as electrical work, in a way that can earn him substantial savings.

While this is an opportunity that most contractors brag about it, the fact is that only few take full advantage of it. Most contractors do not fully implement such a policy because of lack of confidence in their finished product. Therefore, they skim the surface or rely exclusively on their field lead-men's foresight. With this scattered approach, the savings seldom amount to more than changes for which the customer is entitled to the most credit.

A productive redesign reduces the cost without affecting the scope of work or violating any construction codes. This will create a savings of which the contractor is the only beneficiary simply because he is carrying out the work per the terms of his agreement and the industry standards, which generally state: "A job should be constructed in a workmanlike manner, and all conduit-runs should be parallel and perpendicular to building lines and not as shown on the drawings."

Before we address job redesigning itself, two very important thoughts should be considered:

- The only time a contractor can substantially save on something is when that something exists to the fullest. In our case, that something is the total labor and material that will be spent in completing a job. A contractor should be decisive. The sooner he starts saving, the more there is to save on. To start the process at the tail end of a job is futile, for what's left is not enough to make a dent in the costs. Redesigning is a well of opportunities; a contractor should dip in while it's full.

- A contractor should not rely on others to do the saving for him. Delegating this task to others is no different than relying on others' foresight to guide him into success. If a contractor needs to gain time and confidence, he should make a distinction between designing and drawing. He can quickly lay out ideas and design work, even on a brown paper bag if he has to, and later on, if need be, have it drafted neatly.

Who is best suited for laying out and redesigning a job? The office or the field? It's a controversy that can only be settled by emphatically stating: both. For a team there is never a better time to pull together than in the redesigning stage of a job. Here is where each team player can contribute his or her expertise to the common cause of the company—making the most profit on each job. The office will always be subject to the field's scrutiny since it is the field that finally determines the true scope of work. The office, however, is where it all has to start.

In laying out and redesigning a job for the purposes outlined in this chapter the most common items to consider are:

- Benchmarks and survey
- Exposed or slab work
- Trenching and back-filling
- Concrete cutting and core boring
- Feeder layout and conduit fill
- Circuit layout and conduit fill
- Conduit routing and fastening methods
- Device locations and heights
- Lighting fixture pre-assembling
- Lighting fixture support methods
- Control wiring and diagrams

Due to the remote and, most often, liberal design criteria, these typical items are fountains of opportunity for the expert electrical contractor. He should take advantage of his skills and all the National Electrical Code rules and rework all he can in his favor. The mere research for a code rule or a pay-line definition for a specific item can help him boost his profit line considerably.

To further boost that profit line, nothing will stimulate the field into a higher productive mode than the contractor's genuine concern and support for their everyday detail work. This support, when given with authority through job layouts and the like, will prevent confusions and unnecessary downtime, as well as radiate confidence throughout the organization. Every minute spent on the drawing board will increase the job productivity many times over.

A contractor should hold on to his profits. He should apply all his skills to the layout of each item because this may be where his only profit lies.

LABOR

To some contractors, making a profit means trying to make a $10-per-hour man produce like a $20-per-hour mechanic. Without being this extreme, a contractor can drastically increase labor productivity by motivating labor and by assembling the crews that are the most productive for a specific task. For example, installing lighting fixtures and devices with a well-trained crew of $10-per-hour men led by an experienced lead-man is more cost-effective than trying to do the same job with a group of disorganized and overqualified mechanics.

To keep a balance between estimated and actual labor costs, a contractor must always keep in proper perspective the estimated labor rate, for that is what got him the job. In other words, he cannot estimate a job at an average labor rate of, say, $15.00 per hour and maintain his anticipated profit if he actually pays $18 per hour. Nor will his job be profitable if his productivity ratio is lower than estimated. The equation is simple. Labor cost is the product of labor rate times man-hours. To maintain or increase profits, we can only maintain or reduce labor cost by maintaining or decreasing labor rate or man-hours or both. Beware, because every job is bound by this equation; it's entirely up to the contractor to police it.

Manpower needs to be dispatched prudently and effectively. The easiest way to kill productivity is to send a disorganized crew to a job for a couple of hours at a time just for show—a practice disorganized shops are afflicted with when they are overwhelmed with cheap work.

In accepting a job a contractor has taken on the responsibility of completing the job on time and in a businesslike manner. And as a responsible contractor, he should monitor job progress and schedule his manpower (and materials) accordingly. The advantages of applying this principle can be contrasted to the disadvantages of not applying it as follows:

Advantages

- On schedule with job progress
- Total control over the make-up of productive crews
- Confident manpower
- High productivity
- Able to use leftover manpower from other jobs
- No overtime
- Great business relationship

Disadvantages

- Behind schedule
- Disorganized manpower
- Low morale and confused manpower
- Low productivity
- Forced into overtime without compensation
- Poor business relationship
- Responsible for liquidated damages

If business is slow, the contractor's best hedge against falling into a vulnerable position on a new job is to stay ahead of schedule by doing all that can be done at the start of the job. However, if the shop is busy making money on other jobs, then the contractor's best bet is to blitz the job as soon as possible even if it means hiring extra help. Procrastination will lead a contractor into disruption, not only of the new job, but of other jobs as well. The best policy, ultimately, is not to take more work than he can handle in the first place.

Labor comes with moods, time clocks, and a variety of personal problems that a contractor has to learn to respect and to deal with. Because labor responds to moods and attitudes, labor is easy to control. However, disregard man's need for acceptance and recognition and you will lay the groundwork for an explosive situation.

In making up productive crews, besides coordinating work experience and expertise, a contractor must consider the human elements of each crew. Put two veteran electricians—for that matter, any two men—who don't get along in the same crew, and no matter how hard they try, all their experience and expertise will not prevent them from disrupting that crew. On the other hand, place the same two fellows in different crews, and each will increase his crew's productivity.

Manpower is most productive when the relationship is straightforward and each man is part of the team. Telling a worker he did a good job when he did not, just to make him feel part of the team or not to hurt his feelings, is deceptive not only to the team but to the man himself. Taking the worker aside, however, and honestly telling him our expectations and assisting him through the reasoning process is a successful and productive approach that is good for all parties concerned.

The old adage should read: *"How you use labor can make you or break you."* When we least expect it a bombshell can explode and we are back at square one looking for "good" key men. The best way to prevent this kind of explosion is not to lay the groundwork for it in the first place. A contractor should talk straight and never promise what he is not able or not willing to deliver. For more on this subject and on job control, see Chapter 14.

MATERIAL

The third item on our list, and perhaps the simplest to deal with, is material. However, when quantities bought are greater than quantities installed, the subject needs and deserves as much attention as the others.

The materials we're most concerned with here are the basic pipe and wire and all related fittings, fasteners, boxes, plates and terminating devices that generally abound in flea markets and garage sales due to our negligence at adequately policing acquisitions and disposal of surpluses.

To minimize surpluses we must understand the purposes of and the difference between job estimate takeoff quantities and field takeoff quantities.

The objective of compiling quantities during estimating is directed more at generating a bill of material that is in keeping with standard estimating practices, bidding documents, and competition than to the job's ultimate design and need. At this stage we have to be conservative, for doing it any other way is too risky. This principle is mainly responsible for the often vast difference between estimate quantities and field quantities—a difference a contractor must learn to recognize and to capitalize on, for it not only includes the cost of material but its related man-hours as well.

Field takeoff quantities, however, when compiled from drawings and layouts that reflect the latest revisions, are closer to the actual quantities needed to produce a profitable job.

And, because our industry codes and numerous catalogued parts offer our mechanics so many ways to get electrical installations done, it is unwise for the office to arbitrarily order quantities of wiring materials without the direct input of the installers. In fact, the original takeoff sheets should be kept off-limits to the buyer.

Disposal of surplus material, while important, is secondary to acquisition. Buying more material than a contractor needs is a poor investment, as a surplus rarely pays back more than 20 cents on the dollar. In addition, it invites people to steal.

Buying excessive quantities for the sake of saving a few points is, in most cases, not as economical as it seems. Buying a box of 12 or a box of 100 of some special item when a contractor only needs a few, just because the counterman says it's cheaper, is like listening to a bricklayer telling him what stock to buy on Wall Street. When a good opportunity comes along, a contractor has to be sure to include handling and warehousing costs plus pilferage in his savings computation.

Another item to consider with material is cash flow. For a contractor to keep his cash flow healthy a job should be fed periodically and only to the extent that materials are needed for that period. Whenever possible, a contractor should synchronize material releases, including special equipment, with requisition, payment schedules, and suppliers' billing cycles. For example, receiving an order of material on the 28th of the month when the requisitions cutoff date is the 25th will delay payment for that material for approximately 60 days. However, if a contractor arranges its delivery for three days earlier, he can get paid within 30 days and meet his suppliers' due date.

Good material management is essential for keeping job costs down, progress smooth, and cash flow healthy.

TOOLS

"Give me a point of leverage and I'll raise the world," was said long ago. Tools have been on man's mind since the beginning of time. Those who acquire them will always prosper, for they can accomplish otherwise impossible tasks in record time.

In doing a job, as much as we long for higher productivity, safety has to take priority over productivity. Keeping this objective in proper perspective we can still maintain a high, and most often a higher level of productivity if we implement a safety policy that will govern maintenance and distribution of power tools in a timely way and only to those qualified to operate them safely.

With this commitment to safety toward ourselves and others we can now explore the benefits of tooling up a job properly.

Powerful wire-pulling systems, pipe-benders, rolling scaffolds, hydraulic lifters, trenchers, electric saws, crimping tools, knock out sets; stud punches, cable cutters, circuit tracers, Amprobe, step ladders and extension ladders, and the like are part of our work force. The important role these tools play in carrying out the work is most often taken for granted until one tool breaks down or it's missing from the toolbox.

The tools necessary to do a job, like labor and materials, need to be laid out well in advance. Is it more economical to buy or rent? To wait till the last second to repair that electric drill that broke down a few months ago can prove costly today. Often, all the willingness and momentum gained by the manpower can be lost simply because of a wrong-sized tool or a dull drill bit. Now the tool no longer aids but hinders, for it frustrates rather than helps manpower.

To further emphasize the importance of tools, a contractor must remember that estimated labor rates, including those he has developed, are based on availability of materials and tools ready for the installation of work at the job site and not at a supply house or a rental place. The very job a contractor is now doing is also based on that principle. In other words, the contractor's job profit is dependent on his ability to supply and coordinate labor, material, and tools in such a manner that each enhances the performance of the others, thus reducing, or at least maintaining, anticipated job costs.

Mobilizing a job properly is crucial to its outcome. The material covered in this chapter is applicable to most jobs, large and small. Not applying these principles will prove costly sooner or later.

Chapter 14

CONTROLLING THE JOB

CONTROL

Controlling job expenditure is a task that often defeats the best of us. In this chapter we will cover the most common measures a contractor can take to protect the job's estimated costs (labor, material, and direct job expenses) and prevent those unforeseen events that are most responsible for unexpected losses.

Control can be broken down into two major categories:

- Cost Control
- Job Control

COST CONTROL

For our purposes, cost control includes the management and control of expenditures for labor, material, and direct job expenses.

LABOR

Of the three, labor can be the most complex to control when mismanaged. However, when properly managed, it is the only element that can be motivated into a highly productive mode. Therefore, a contractor should learn and employ methods that will help him accomplish that.

If a contractor wants labor to produce up to his expectations, then he must let labor know what his expectations are. The contractor should provide his lead-men with tools that will help unleash their knowledge and desire to produce. For, all in all, given the opportunity, people will always strive to excel.

One way of doing this is to give the lead-men a road map showing where the contractor is coming from and where he intends to go. Clearly, a contractor should delegate to them the responsibility of taking him to his destination safely. In this equation, "where the contractor is coming from" is his estimated man-hours, and "where he intends to go" is the allocated man-hours, with "his destination" being the actual man-hours spent in getting there.

The simplest method of conveying estimated, allocated, and actual costs to a job lead-man is in: man-hours for labor, quantities for material, and lump sum for direct job expense. To expect a lead-man to work with an array of computed ledger sheets reflecting extended labor and material costs or other data that go beyond the scope of controlling and managing labor is an imposition that unnecessarily overburdens him with paperwork.

Providing a lead-man with a concise and tangible method for tracking estimated and allocated man-hours versus actual man-hours spent communicates the contractor's expectations to him in a straightforward manner that is welcomed by most lead-men and gives them a road map of how to manage the job itself. When a contractor does this he is also demonstrating positive leadership which ultimately radiates throughout his entire labor force—a domino effect, if you will. The contractor will instill self-esteem that enhances pride and responsibility in everyone who is involved. By giving the lead-man that road map, the contractor is giving him a sense of purpose and direction that stays with him 24 hours a day.

In our cost control method, the setting of this road map is done openly and jointly with the lead-man, for it sets strong foundations for a productive labor relationship. For example, the "Labor Master Control Sheet" will show the estimated hours-breakdown. Jointly the contractor and his lead-man, for the first time, will evaluate and lay out the different crew types and agree on the number of hours he feels are really needed to do the work both effectively and efficiently. The agreed hours and crew types are then posted in the "Allocated Hours" column.

A contractor should not underestimate this process because it offers him the only opportunity to set a realistic production goal that is endorsed by his lead-man at the start of the job—an endorsement that most lead-men will take pride in delivering on.

This control method, when laid out as shown, will also assist the lead-man in pinpointing his crews' production rates, thus preventing setbacks. For example, to monitor the installation of a 1000 ft. of 4" EMT conduit run for which a contractor allocated 100 man-hours and a two-man crew at anytime during the installation, simply compare the production output to the expected rate—in this case two lengths per crew-hour. For instance, at the end of the second day if the contractor doesn't see approximately 320 ft. of conduit in place, he knows his production is off target and needs immediate attention.

This production rate, or reference point, can also be established for any item of work listed in the job breakdown. It can be monitored by anyone concerned, including a member of the installation crew itself.

LABOR INCENTIVE

If a contractor wishes to set an incentive program for the lead-men, or for the entire labor force, for that matter, this method is realistic and self-motivating. Realistic because it deals with tangible numbers that everyone can relate to and specifically rewards a lead-man for what he was hired to do, manage and control labor; and self-motivating because its outcome is independent from the company's overall performance. In other words, regardless of the job's profits or losses, he's assured a flat rate bonus for every hour he saves from the job's total estimated man-hours.

Optionally, to keep the flame burning, using good judgment and an overall job production rate as a guideline, a contractor may allow his lead-man to draw a monthly percentage of his anticipated share, which is the difference between estimated and actual man-hours times the per-hour bonus. To avoid misunderstandings, a contractor must agree to this per-hour rate before he implements the program.

The hourly rate used in computing bonuses should not be confused with the average labor rate used in estimating the job. Here the objective is to simplify bookkeeping in establishing a dollar rate he will pay for each hour saved. For example, a contractor may agree to pay an arbitrary rate anywhere between $3 and $14.00 per hour or he many base it on a percentage such as 25%, 50% or more of the job's average labor rate cost. Whichever method he uses in determining this rate, the contractor should remember this incentive program is strictly based on man-hours saved and not on overall job profits.

This method saves time because of its simple approach which requires no searching of unrelated job records. To compute the bonus earned, neither the contractor nor his lead-man need examine complex records and confidential reports; nor need the lead-man wonder whether he will be credited the full amount earned, for he will be doing the recording.

LABOR CONTROL SYSTEM

Besides enhancing productivity, this control system gives the lead-men an indispensable tool that will help manage and report the job progress and other pertinent field-generated information, such as completed percentages of individual items of work or change orders, in one concise report—the "Weekly Labor Report Sheet." This form, along with the "Daily Assignment & Man-hours Report" sheet, provides the contractor with the detailed information necessary to compile reports such as weekly payroll, payment requisitions, labor productivity, job cost analyses or any other report that can be supported by this data.

The system consists of:

- Labor - Master Control Sheet
- Daily Assignment & Time Card
- Weekly Labor Progress Report

Before a contractor begins reading the explanatory notes to these three forms, he should remove each form from the manual and hold it adjacent to his reading. This method will increase his appreciation and understanding by helping him compare notes quickly while stimulating his thoughts as to the various ways in which the data may be applied to his specific situation.

LABOR MASTER CONTROL SHEET

See Figure 9 and Figure 10. This sheet is where all data pertinent to a job's production and to its incentive program is gathered so as to manage and maintain the primary objective—making a profit on each job.

To save time and prevent human error in recording numbers a contractor should make a template by filling in all constant information, such as the top three lines, exclusive of the date, and Columns 1, 2, 3, and 4. Once the template is completed the contractor should make a few copies and use them as worksheets. Anytime a datum becomes constant, you should modify the template accordingly and make new worksheets. An example of a constant datum is an item of work or a change order that is 100%

complete.

The form has two sides: basic contract and change orders. Both sides are divided into three main sections: estimate, production, and incentive. With the exception of the estimate section, the two sides are similar.

ESTIMATE SECTION - Basic Contract

Columns No. 1 & 2 (No. & Items of Work)- Items of work and system numbers should be consistent throughout the paperwork. For the sake of continuity, in our example we are carrying over our "Sample Job 1" data right from its "Requisition No. 2" (see Chapter 12).

Column No. 3 (Estimate Crew - Hours)- should reflect the crew types and the estimated hours used to estimate the job.

ESTIMATE SECTION - Change Orders

The back side of this form is laid out to handle all the change orders. At the top of the page, it provides a space in which to log "Sheet No. __of __." If the number of change orders exceeds the number of lines, simply attach another sheet (a contractor may photocopy the "Change Order" side of the form) to his Master Control Sheet and identify it as Sheet No. 2 and follow the instruction at the bottom of Sheet No. 1.

A typical form can handle up to 24 change orders. When we exceed that number, we simply create sheet No. 2.

Column No. 1- Post each change order number here, regardless of whether it is pending or approved.

Column No. 2- Change order title or a brief description

Column No. 3- This column, when properly used, will keep the change order status up-to-date. When a change order, for whatever reason, is canceled, never delete it from this list; we simply post a code letter, such as "R" for Rejected or "H" for Hold, in the Pending column. A change order submission, especially the one we're generating, is a kind of report and a warning to the customer of how things should be done and why. This method will give a contractor a valuable sequential record of his change order submissions and will give continuity to a sequential numbering system.

Pending and Approved Columns- When a change order is pending, post the estimated man-hours in the pending column, and when it is approved we simply post the approved number of hours in the Approved column and allocate a crew type and man-hours in Column 4.

PRODUCTION SECTION

Column No. 4 (Allocated Crew - Hours)- This is where a contractor sets his goal. The contractor posts the number of man-hours in this column that he and his lead-man anticipate he will need for each item of work. There will be times when the reality of the situation forces a higher number of man-hours than a contractor has estimated, no matter how hard he tries to stay within that estimate.

When this occurs, if a contractor wants to save the day, he must be realistic and straightforward with his lead-man and accept two basic facts: (1) the lead-man did not estimate the job, and (2) any deception on his part ultimately can only be counter-productive. The best a contractor can do at this juncture is to analyze all possibilities and post the best number they both think they can do that item of work for. Hopefully, the job as a whole will average out to a profit.

Column No. 5 (Actual Used Hours - %)- Actual hours used are those reported under Column No. 8 of the "Weekly Labor Progress Report." Because Column No. 8 is cumulative, that is, the last report has the latest total of hours used, anytime we post these hours in Column No. 5 it updates the master control sheet. In other words, we need not update the master control sheet weekly. We may update it at requisition time or at any other time we wish by simply posting the latest entries shown in the "Weekly Progress Report."

To compute the percentage (%) of man-hours used in Column No. 5, we simply divide actual hours used (Column No. 5) by allocated hours (Column No. 4) times 100.

Column No. 6 (Work Done %)- Percentage of work done is recorded in Column No. 10 of the "Weekly Labor Progress Report," which is the lead-man's estimation of the work complete when he hands in his report. These percentages, however, because they are ultimately used for requisition for payment and to monitor the job progress and the incentive program, should be field-verified and adjusted as often as necessary.

INCENTIVE SECTION

Column No. 7 (Earned Hours)- Earned hours are the number of hours the lead-man or crew has earned under the incentive program. The hours earned under this column are computed by deducting actual hours used from estimated or allocated hours, whichever amount is larger. The final earned hours on each item of work can only be computed when the item is 100% complete.

Column No. 8 & 9 (Paid & Balance Hours)- will track, in hours, any hours paid as bonus and balances to be earned on each item of work.

Red Flags- In both examples, Figure 9 and Figure 10, we will note shaded lines. We may use a highlighter. These are red flags telling us, for example, to watch out for Items No. 11b and 14 because their allocated hours are greater than their estimated hours. Any situation that indicates concern should be red-flagged and monitored until the item is out of danger or 100% complete.

Analysis- How a contractor interprets the numbers shown on the bottom line of this form has a lot to do with his perception of the system as a whole. Simply posting and tracking hours for the sake of knowing where a contractor is at any given time has a certain value. However, being able to see red flags in time and prevent losses while he has the chance to recover is what this system is all about. It can only be classified as invaluable. The contractor's knowledge and imagination in using these numbers may well become his key to success, for they also offer a solid foundation for his next estimate.

LABOR MASTER CONTROL SHEET								
Base Contract								

Job Name: Sample job 1, Warehouse   Job #02-1231   Ending Period: 2/28/___

Lead-Person: Allan Smith   Incentive Rate: $7.50 Per Hour   Approved by: LS   Date: 2/28/___

(1)	(2)	(3) Estimate		(4) Allocated		(5) Actual Used		(6)	(7)	(8)	(9)
								% of work done			
No	Item of Work	Crew type	Hours	Crew type	Hours	Hours	%		Earned hours	Paid hours	Balance hours
1	Mobilization (DJE)										
9	Service	II	49.30	II	36.00	28.00	57	80			
10	Secondary Feeders	II	68.68	II	54.00	40.00	58	80			
11a	Switch-gear on site	II	2.00	IV	1.00	1.00	50	80			
11b	Switch-gear install rough	II	11.35	II	16.00	13.00	81	80			
11c	Switch-gear install final	II	4.00	II	2.00						
12a	Branch circuits conduit	II	59.82	II	48.00	28.00	47	60			
12b	Branch circuits wire	II	25.92	II	16.00	5.00	19	30			
13a	Lighting fixtures on site	IV	6.00	IV	3.00						
13b	Lighting fixture installation	IV	48.50	IV	42.00						
14	Wiring devices	IV	14.03	IV	16.00						
!5	Equipment connections	II	4.87	II	6.00	2.00	33	20			
20	Fire alarm	II	57.44	II	40.00						
	(This and other forms may be downloaded online at www.theestimatingroom.com)										
	Base Contract Total		351.91		280.00	123.00	35	46			
	(From Other Side) C.O. Total		11.00		11.00	6.00	54	60			
	Revised Contract Total		362.91		291.00	129.00	36	47			

Form EC-401-rev05 ©2005 The Estimating Room™ Inc.

Figure 9: Labor Master Control Sheet- Base Contract

LABOR MASTER CONTROL SHEET								
Change Orders Sheet # 1 of ____								
Job Name: Sample job 1, Warehouse			Job #*02-1231*			Ending Period: *2/28/__*		
Lead-Person: Allan Smith		Incentive Rate: *$7.50* Per Hour			Approved by: LS	Date: 2/28/20___		
Estimate			Production				Incentive	
(1)	(2)	(3) C.O. Status	(4) Allocated	(5) Actual Used		(6) % of work done	(7)	(8)	(9)
No	Change Orders	Pending	Approved	Crew type	Hours	Hours	%		Earned hours	Paid hours	Balance hours
1	Add interior lights		3.00	IV	3.00	1.00	33	40			
2	Wire irrigation pump		10.00	IV	10.00	7.00	70	80			
3	Relocate power devices	6.00									
4	Delete 2 outlets		-2.00	IV	-2.00	-2.00	100	100			
(Full size template of this form may be downloaded online at www.theestimatingroom.com)											
Total this sheet (No. 1)	6.00	11.00		11.00	6.00	54	60			
Total from sheet No. 2										
(Transfer to other side) Grand Total	6.00	11.00		11.00	6.00	54	60			

Form EC-401-rev05 © 2005 The Estimating Room™ Inc.

Figure 10: Labor Master Control Sheet - Change Orders

WEEKLY LABOR PROGRESS REPORT

Typically, this report is laid out on both sides of the page (see Figure 11 and Figure 12). The front is for the Basic Contract, and the back is for change orders. Its data is generated from the "Daily Assignment" sheet, (Figure 13) and, when computed, its totals may be transferred to the "Master Control Sheets."

Columns No. 1, 2, 3 & 4 (Items of Work)- This is constant data that has to synchronize with the master control sheets of this system; in fact, we should copy it from those sheets. Once these columns are filled in on both sides of the page, use the original as a template to generate more worksheets.

Column No. 5 (Hours Used As Of Last Report)- In our first report the value of this column is zero. In all subsequent reports, its value is that of Column No. 8 of the previous report.

Column No. 6 (This Report)- The information for this column is generated from the daily assignment sheets—one for every day worked on that job. For better recordkeeping, we should post the day's date in the blank spaces below the weekday.

Column No. 10 (% of Work Completed)- This is the lead-man's estimate of the work completed, up to and including this report. These percentages are essential to complete the payment requisitions and to monitor the incentive program. We should verify them at least once per billing period.

Change Orders Side- As with all other forms presented in this manual, the change orders side of these forms can be extended to as many sheets as needed.

WEEKLY LABOR-PROGRESS REPORT

Base Contract

Job Name:		Job #		Week-ending:
Lead-person		Checked and Posted by:		Date:

(1)	(2)	(3)	(4)	(5)	(6)								(7)	(8)	(9)	(10)
					Actual Hours Used										Balances	
				As of	This report										Hours to	% Work
			Allocated	last	S	M	T	W	T	F	S	Total	To date	complete	Complete d
	Items of work	Crew	hours	report								hours	(5 + 7)	(4 - 8)	to date

(This and other forms may be downloaded online at www.theestimatingroom.com)

	Base contract total															
	(From Other Side) C.O. total															
	Revised contract total															

Form EC-405-rev05 ©2005 The Estimating Room™ Inc.

Figure 11: Weekly Labor-Progress Report - Base Contract

WEEKLY LABOR-PROGRESS REPORT															
Approved Change Orders													Sheet # of		
Job Name:						Job #							Week-ending:		
Lead-person						Checked and Posted by:						Date:			
				Actual Hours Used									Balances		
(1)	(2)	(3)	(4)	(5)	(6)							(7)	(8)	(9)	(10)
				As of last report	This report							Total hours	To date (5 + 7)	Hours to complete (4 - 8)	% Work Completed to date
			Allocated		S	M	T	W	T	F	S				
	Items of work	Crew	hours												
	(Full size form may be downloaded online at www.theestimatingroom.com)														
	Total this sheet (No. 1)														
	Total from sheet No. 2														
	(Transfer to other side) G total														

Form EC-405-rev05 ©2005 The Estimating Room™ Inc.

Figure 12: Weekly Labor-Progress Report - Change Orders

DAILY ASSIGNMENT & LABOR HOURS REPORT

This form, besides helping the lead-man every day in assigning the work to his manpower, is the basis for the labor control system and an integral part of the job diary, for it offers detailed information on who worked where and when. In essence, it makes both systems (Labor Control and Diary) easy to maintain and track.

Change Orders- Side is not shown, for it works on the same principle as it does in other forms.

Template- Once we post the constant data, we should keep the original as a template and generate as many worksheets as we need.

Crew I.D.- The crew type used in the master control and weekly progress reports should correspond to this line, for that is the goal that the contractor and his lead-man set at the start of the job. Any deviation should be carefully evaluated if we want to stay on target toward our goal.

Shop, Job, Single System- The third line down from the top gives us a choice of using this form in one of three different ways:

Shop- A contractor may use it as the daily schedule for his entire shop, in which case the items of work are his jobs or work orders and his crew **I.D. may be his truck number. Job.** This will indicate that the form is used in conjunction with the labor control system, listing the original contract breakdown as items of work as laid out in the master control sheets (Figure 9 and Figure 10).

Single System- For better control in some large projects, it is advantageous to further break down individual systems into smaller increments of work. In such a case we would use a sheet for each of those systems, or we may use a sheet for each crew.

The function of this sheet is to accurately keep track of hours spent in completing an item of work or an entire job. Its records are essential to any successful operation.

DAILY ASSIGNMENT & MAN-HOURS REPORT															
Job Name:					Job #						Lead-person:				
( )Shop ( )Job ( )Single System:					Day of Week:					Date:					
Base Contract			Daily Production Schedule												
	Worker's Name or I.D. #														
Items of work	Crew Type→														Hours total
(This and other forms may be downloaded online at www.theestimatingroom.com)															
This sheet total															
(From other side) C.O. Total															
Today's grand total															

Form EC-401-1-rev05 © 2005 The Estimating Room™ Inc.

Figure 13: Daily Assignment & Man-Hours Report

MATERIAL

As with any other expenditure, material needs to be managed or its cost will run out of control. A common mistake is delegating its entire management to the job lead-man. While the lead-man can and should control quantities ordered and installed, the buyer has the responsibility to control material costs.

Buying power is a direct function of the contractor's relationship with his suppliers and the industry. If he discounts his bills, not only will he pick up a few extra points on their face value, but most likely he will get the preferred customer's discount, not to mention extra consideration when he's quoted special equipment packages.

In this section, however, we shall present a cost control method for material that, like the one for labor, will keep the job operation smooth and profitable.

The first objective in controlling a job is to open the channels of communication between the buyer and the field lead-man, for the interaction of their functions is essential. For example, the lead-man makes up a bill of material for a specific phase of work, and the buyer, after the acquisition, tells the lead-man what was bought from whom and when to expect it. What should be an easy exchange of critical information between office and field can become costly if the communication is poor or non-existent.

The form shown in, "Bill of Material," is used for that purpose. The purpose of the form is to allow the user to list all materials and tools needed to perform a task and submit them for acquisition, and to allow the buyer to readily place orders with the various suppliers. Thus the buyer records, for future follow-up by the field, the suppliers' names, purchase order numbers, delivery dates and prices on the original bill of material.

BILL OF MATERIAL

Job Name: Sample job 1, Warehouse   Job #02-1231

(x)Job ( )Single System: Prepared by: MS   Date: 1/20/20__

Purchase Order #	Ordered By	Source Code	Vendor	Contact Person	Contact Person Phone #
2345-021231	John	A	All Star Electric Supply	Mark	(123) 444-5555
2355-021231	John	B	Sunset Electric	Adams	(123) 356-4545
2358-021231	Joe	C	ABC Rental	Jim	(123) 345-2234
		D			

Codes: **A** through **D** = Suppliers—Alternate sources **S** = Shop **J**= Job Site **O**= Other job sites **X**= Purchase-order

	Ordered	Recv'd	Bk-order	Source Code	Materials/Tools	Delivery Date	Price	Per
1	1200			A	½" EMT	1/24	14.75	C
2	380			A	¾" EMT	1/24	24.10	C
3	100			S	½" EMT Couplings DC SS			
4	36			S	¾" EMT Couplings DC SS			
5	8			A	1" EMT Couplings DC SS	1/24	0.74	E
6	128			S	½" EMT Connectors DC SS			
7	6			S	¾" EMT Connectors DC SS			
8	2			A	1" EMT Connectors DC SS	1/24	0.68	E
9	125			S	½" EMT Straps			
10	250			B	½" Caddy Clips	1/28	0.40	E
11	250			B	¾" Caddy Clips	1/28	0.41	E
12	18			B	¾" EMT Beam Camps	1/28	0.41	E
13	2			A	½" to ¾" J,B. Combo	1/24	1.45	E
14	10			B	4" Square Box w/cover	1/28	1.60	E
15	34			B	4" Round Box w/cover	1/28	1.10	E
16	1			O	2 Section -6ft Rolling scaffold	Pickup		
17	1			C	Rotating Hammer	1/26	16.00	Day
18	2			X	Back boxes for Panels A & B	1/26/	Fixed	E

Form EC-430-rev05 ©2005 The Estimating Room™ Inc.

Figure 1: Bill of Material

MATERIAL CONTROL SYSTEM

Material control, like labor, offers its own challenge. The initial and best approach for preventing the inevitable overbuying of standard items is to keep close scrutiny of the supplier's invoices as they come in. If necessary, a contractor should insist that his supplier bills him daily.

The second best approach is to discard the estimated quantities used to bid the job and generate bills of material reflecting the actual job conditions and latest revisions. As discussed earlier in Chapter 13, *"Re-designing the Work,"* using the original estimate to buy standard material will be counter-productive to that objective.

The form that follows is a part of the overall cost control system. In most jobs it will facilitate tracking and recordkeeping of estimated material versus used material.

MATERIAL COST CONTROL

Job Name: _____   Job #: _____   Prepared by: _____   Date: _____   Period Ending: _/_/_

Items of Work	ESTIMATED COST — MATERIAL			AS OF LAST REPORT			ACTUAL PRIME COST — THIS REPORT			TOTAL TO DATE		
	Standard	Quote	Total	Standard	Quote	Total	Standard	Quote	Total	Standard	Quote	Total
Base Contract Total Material Cost:												
(From Other Side) C.O. Material Cost:												
Grand Total Material Cost:												

(This and other forms may be downloaded online at www.theestimatingroom.com)

Form EC-440-1-rev05 ©2005 The Estimating Room™ Inc.

Figure 15: Material Cost Control

JOB COST CONTROL

Regardless of whether we use an electronic spreadsheet or a conventional system similar to the one presented here, if we want to control the job cost effectively, we must begin the process at the start of the job. To re-emphasize this point we shall refer to a specific quotation from Chapter 13 "Redesigning the Work":

The only time we can substantially save on something is when that something exists to the fullest. In our case, that something is the total labor and material that will be spent in completing a job. A contractor should be decisive. The sooner he starts saving, the more there is to save on. To start the process at the tail end of a job is futile, for what's left is not enough to make a dent in his costs. Redesigning is a well of opportunities; a contractor should dip in while it's full.

The job cost control system—in fact, all cost control systems presented in this manual—can be converted into electronic spreadsheets. If the contractor is well versed in the use of electronic spreadsheets such as Microsoft Excel, Lotus 1, 2, 3, or any other program, once he has learned the system shown here, he's encouraged to make that conversion. The following form (Figure 16) is a prime candidate for such a conversion.

The Form- The form is basically divided into two sections, "Estimated" and "Actual Cost," with the front page reflecting the basic contract and the rear change orders.

Labor is reported in dollar amounts rather than in hours, and material includes DJE. Markups and anticipated profits are not posted here. This sheet deals in prime costs only.

( ) A Job ( ) All Jobs- The form may be used for a single job, in which case we should list the standard "Items of Work," or for several jobs, in which case we would list all the jobs, one job per line.

We should bear in mind when using this form for all jobs that we still have to summarize each job on its own sheet before we can list it on this sheet. The "All Jobs" sheet will give us an overview of the overall performance of all the active jobs.

Estimate Cost- These are the estimated costs from the "Estimate Recap Sheet."

Actual Prime Cost- The labor cost reported here is the gross payroll cost plus direct labor burden. We may choose to multiply used man-hours times the labor rate.

		JOB COST CONTROL						For ( )a job ( )all jobs		

| Job Name: | | | | Job # | | Prepared by: | | Date: | | Period Ending / / | |

Items of Work	Job(s) ESTIMATED COST			AS OF LAST REPORT			ACTUAL PRIME COST THIS REPORT			TOTAL TO DATE		
	Labor	Material	Total	Labor	Material	Total	Labor	Material	Total	Labor	Material	Total

(This and other forms may be downloaded online at www.theestimatingroom.com)

Base Contract Total:

(From Other Side) C.O. Total:

Grand Total:

Form EC-460-rev05 ©2005 The Estimating Room™ Inc.

Figure 16: Job Cost Control

Conclusion to Job Cost Control Systems

The most effective cost controls are those in which the principal parties participate in maintaining them. When implementing a new system, we should remember that all participants are human, and by nature they will oppose anything that threatens the way they are accustomed to doing things, especially when they are not aware of the overall benefit to the organization.

The cost control system shown in this chapter is designed to involve and educate as many participants as possible. This system will make the best use of each person's contribution. For example, the lead-man manages labor, the buyer manages material, bookkeeping converts both reports into the job cost control sheets, and the contractor, with facts and figures in hand, can make periodic field inspections to verify job progress and overall performance.

In other words, everyone is contributing something of value to the system while the paperwork provides the contractor a reliable checks and balances method.

If we choose, as we should, to adopt a control system, then we must educate our people and police them until the system becomes second nature to them; for, as stated earlier, it will most likely meet with stiff opposition and all sorts of reasons will be given why we should abandon or simplify it. A contractor should stand by his convictions, and in the long run it will pay him handsomely.

Job Control

Job control—as opposed to "Job Cost Control," which is based on hard figures—offers another way of controlling the outcome of a job. It is based on what experience has taught us. In that spirit, we present the following "Job Control Tips" from some successful contractors:

Alteration Work- A contractor should not allow others to do work under his permit—for example, the landlord who wants his handyman to install a couple of exits and emergency lights down the hall. Besides being legally responsible for the work, a contractor is giving up work that belongs to him or some other licensed contractor.

Time and Material Work- A contractor should not allow the customer to assist him with his manpower. Aside from the issue of legality and responsibility, in T & M work the use of another's manpower, unless it goes through the contractor's payroll and billing system, cuts into his profit.

Field Wiring of Equipment and Machinery- Before a contractor starts field wiring equipment or machinery, he should make sure the customer supplies him with workable and complete electrical diagrams as needed. The contractor should analyze them well and keep copies for his files. His work is to wire electrical apparatus according to sets of workable drawings and not to lay out and design logistic diagrams. For lack of understanding of this relationship, too many contractors cross their line of responsibility. What begins as goodwill usually ends up in costly disputes and disappointments.

When to Start the Job- A contractor should not depend on others to tell him when the job is ready for him to start. He is the expert electrical contractor, not his customer. Depending on the customer's judgment as to when he should begin a certain phase of work is no different from listening to him when he indiscriminately calls for more manpower. From beginning to end and in between job phases, a contractor should always visit the job and look for new opportunities to advance his work in the most economical way. For example, calling the contractor to rough in a bunch of home runs under a slab that was to remain and now is removed is the last thing on the customer's mind.

Shop Drawings- A contractor should never install material that is not approved for its use. Before a contractor installs any material, he should make sure the shop drawings are approved for the intended use and jibe with his scope of work. Shop drawings that are *"Approved as Noted"* often have subtle changes that escape many reviewers. For example, a stamped blue arrow which points to a specific detail may be lost in a color background or washed out in duplications.

One of the most frequent blunders is a voltage change in lighting fixtures and equipment, making a shambles of an entire installation, especially when a contractor has combined power and lighting circuits in the same raceways.

Photographing the Work- Most contractors fail to take full advantage of Polaroid photographs. A contractor should make it a special task. Recording his work before walls or ceilings are closed may be the only witness to what was there before someone damaged, covered, or relocated his work. It will also help in determining the amount of work completed at a given time and may substantiate or defend future claims or become part of a promotional brochure for the company.

An example of photographs saving a contractor a bundle occurred when a customer wanted to back charge the electrical contractor for redoing a sheetrock ceiling. When the ceiling was completed it became apparent that most of the 400 hi-hats were misaligned. However, photographs taken of the hi-hats in place before the ceiling was installed clearly showed each row of hi-hats in perfect alignment. Apparently, the mechanical contractor accessing the ceiling in several areas to rework ductwork and equipment was the culprit. Thanks to photographs, the electrical contractor proved his case and charged the customer for the extra work.

A contractor should learn from this example and take and date photographs of his work as often as it is needed.

Diary- A diary is like a photograph of the day's events documented in words. Like photographs, a good job diary can prove to be invaluable in a dispute.

Pay-line- That fine line between getting paid and not getting paid for the work we do is here called pay-line. A contractor should not forget that all we get paid for is what we do. Doing work that is not part of our contract is counter-productive. The contractor should pay special attention and clearly maintain a contract overview of his scope of work.

Extra Work- Anything beyond the pay-line is extra work for which a contractor must get paid. A contractor should get the proper documentation and payment method before he commences any extra work. The emphasis here is to stick to the terms of the contract.

Change Orders- Proceeding with an unauthorized change can make a contractor liable for the change itself. Construction documents and scope of work can only be modified by written directives called "Change Orders" issued by the customer.

Doing the job can be fun, in fact, if a contractor applies good working principles and qualified labor, it is not only fun but profitable. And profit is the reason we are in business.

Chapter 15

INSPECTIONS

THE ELECTRICAL INSPECTOR AND THE SYSTEM

The electrical inspector represents the people in his community. His job is to ensure that the work he inspects is done in accordance with all applicable codes and ordinances and no hazardous conditions exist or will arise for the present or future occupants of the premises. Understanding his job will perhaps alleviate the often undue resentment the contractor feels from job delays caused by failed inspections and re-inspection follow-up procedures.

Municipalities, more than ever, are targeted for accountability of faulty installations under the permits they grant. Therefore, they are training for and demanding from their inspectors thorough and comprehensive inspections which, coupled with their mandatory procedures, make the entire inspection process appear intimidating and unwarranted.

Municipalities, by the nature of things, are mandated to protect themselves against the inevitable lawsuit, and in doing so are forced to set extraordinary procedures. However, the frustration that some contractors experience in dealing with their municipalities most often stems from non-compliance to local ordinances and from their own lack of good recordkeeping and follow-up procedures. (This is not to say that some municipal employees don't take their mandate to extremes at times.)

Conventionally, the most frustrated contractor uses a filing method where permit applications and all other related documentation are scattered in job and general administration folders. To stay in tune with the ever increasing regulations and demand for recordkeeping from the various municipalities, the contractor who seeks relief from his frustrations can no longer afford this method.

To cope with this trend a contractor must offset the disparity that exists between the city hall and the contractor in permit and inspection procedures by setting his own procedures that will allow him to keep track of all that is pertinent to him and his job in a specific dedicated filing system—even if it means duplicating records. When a contractor needs accurate information to protect his interests he doesn't want to spend days searching completed job folders hoping to find what he needs.

Most often a well-kept "Municipal Folder" and an "Inspection Log Book" similar to those described below will do the job.

MUNICIPAL FOLDER

A contractor should get a six-section legal-size folder for each municipality he deals with. His investment will be well worth it. If he's dealing with several municipalities, the contractor should allocate an entire filing drawer and label it "Inspections." In each folder he should keep:

Section 1   Copies of all applicable licenses required to register and to pull permits with that municipality.

Section 2   Copies of insurance certificates sent.

Section 3   Active permits issued by that municipality. If needed, he should make a copy for his job file.

Section 4   Inspection reports such as rejections and approvals. If needed, he should make a copy for the job file.

Section 5   General correspondence and yearly renewal applications.

Section 6   Blank permit applications.

INSPECTION LOG BOOK

An important part of our business is to keep good records of all the inspections. The following "Job Inspections Log Sheet" is set up for a three-ring binder to form an inspection log book.

Erasable Writing- In using this form, because the information is bound to change often and because we're human, a contractor should use a pencil or any other erasable writing instrument.

Job Info & Municipal Info- When properly filled out, this provides a contractor with the basic information needed to call in an inspection.

Types of Inspections- The description for each type of inspection (Temp, Slab, Walls, Ceilings, Service, and Final) can be expanded or modified to fit a contractor's needs.

Call # - Each type of inspection can be called in four different times.

P/F- P = Passed inspection; F = Failed inspection.

JOB INSPECTIONS LOG SHEET

Code for P/F Column: P = Passed ; F = Failed
(Write with erasable material)

Inspections Record

Job Info	Municipal Info	Call #	Temp: Date	P/F	Slab: Date	P/F	Walls: Date	P/F	Ceilings: Date	P/F	Service: Date	P/F	Final: Date	P/F
Job #	Permit #	1												
Name:	Inspector:	2												
Addr:	City:	3												
Phone:	Phone:	4												
Job #	Permit #	1												
Name:	Inspector:	2												
Addr-	City:	3												
Phone:	Phone	4												
Job #	Permit #	1												
Name:	Inspector:	2			(This and other forms may be downloaded on line at www.theestimatingroom.com)									
Addr:	City:	3												
Phone:	Phone:	4												
Job #	Permit #	1												
Name:	Inspector:	2												
Addr-	City:	3												
Phone:	Phone:	4												

Job Records in this sheet are from Job # _____ to Job # _____

Check this box when all the jobs in this sheet are completed ( )

Sheet No.

Form EG-150-rev05 ©2005 The Estimating Room™ Inc

Figure 17: Inspection Log Sheet

JOB INSPECTION

Visual impact, as discussed in marketing, goes a long way in marketing and in anything else we do. Presenting a job for inspection is marketing our mechanical skills to those who are in a position to pass or fail our work.

The visual impact our work creates tells the trained eye, and most often the lay person, the type of work we do. For example, if a contractor installed a horizontal 1/2" PVC conduit run to a pool's time-clock with sufficient straps and back straps to keep it from sagging and allow the rain-water and the dirt to seep through between the wall and the conduit and 1/2" liquid-tight to the pump with two mineral-lack straps instead of one, the visual impact he has created builds the inspector's confidence in his work and in him as a conscientious mechanic.

On the other hand, if a contractor presents the same job with straps at three foot intervals with no back straps and one mineral-lack strap on the liquid tight, even though it may be in accordance with the National Electrical Code, it does not pass the unwritten workmanlike-manner code. The work in this case sends a message of a marginal installation at best, prompting an inspector to open and scrutinize the work in detail. Rightfully, inspections progress deeper with each suspicious item the inspector encounters.

The visual impact we refer to here can only be created by those mechanics who are capable of installing quality work above code standards and by those who are dedicated to serving the industry well and want to excel. Therefore, when an inspector encounters this kind of work and passes it with light scrutiny, we can all be assured that his decision did not stem from any deception, but from good judgment.

When you lay out a job, keep those thoughts in mind and provide that extra touch that makes the difference between passing and failing.

Chapter 16

COLLECTIONS

PREVENTION

One of the most talked about subjects among contractors is how to collect money earned on contracts. The prudent contractors make proper arrangements to collect their money before they extend credit. They also never attempt to collect the full contract amount in a lump-sum payment at the completion of the job. They break down the contract amount in installments, each triggered by a due date. Each installment due date is then set by the completion of certain items of work or a periodic cutoff date. Their contract will also have a specific provision for payment and for work stoppage.

An important thing that will aid the collection of money is the right to stop work when not paid on time. Simply put, a contractor cannot stop working on a job because of late payment and shift the liability to the customer, unless the contract stipulates that the contractor can do so and under what conditions.

Typically, most contractors treat this vital matter lightly and go on signing lopsided contracts subjecting themselves to undue frustrations and to possible losses. An example of a lopsided and uncontrollable provision in a contract is when the contractor agrees to get paid by the general contractor only when the general contractor gets paid by the owner, with no specific due dates set and no rights to stop work.

To prevent unpleasant events, a contractor must read his contracts carefully (see Chapter 8) and ensure that a right-to-stop-work clause is inserted in his contract. The clause need not be complex. In fact, clauses are most effective when handwritten on the contract form at the time the contractor negotiates the contract. Such an insertion may read: *"If the customer fails to make a payment within seven days from its due date, then the contractor has the right to stop work until all moneys due are paid in full."* Then if the customer breaks his promise the contractor can stop work without liability. We should remember, handwritten notes on contract forms take precedence over all other printed matter.

The contractor should negotiate his contract, for no reasonable customer will refuse to negotiate. In fact, many expect a contractor to object and to ask for changes. If he doesn't, the customer may grow suspicious, for they know that most other contractors, rightfully, have asked for the same changes he is asking for now.

PROTECTION

Credit Check

The best protection is knowledge and prevention. A contractor should check all his prospective customers for credit references before he extends credit.

Mechanic's Lien

While a Mechanic's Lien is a powerful legal instrument, if we don't follow proper procedures and adhere to filing deadlines that lead us to its enforcement, a Mechanic's Lien can be as worthless as a bad check.

The laws that govern Mechanic's Liens vary from state to state. The suggestions that follow are general and presented to acquaint you with the subject. A contractor should learn the specific procedures that are required by the laws that govern Mechanic's Liens in his state and, whenever possible, use the expert services of a lawyer or that of a local filing firm that specializes in this field.

A contractor must make a distinction between Public and Private Properties. Mechanic's Liens are mostly applicable to private properties owned by private owners where the performance of and payment for the work is not guaranteed by Surety Bonds. An owner is a legal entity and can be natural, corporate, or a partnership. Statutory Bonds are not for subcontractors and should not be confused with Surety Bonds.

The Spirit of the Law

The spirit of the law is to protect those who supply labor and material for the improvement of others' properties and prevent the owner from paying twice for the work done. The general concept is for the general contractor to file a notice of commencement and the subcontractors or suppliers to notify the owner that they have been hired by the general contractor to improve his property. Obviously, if the contract is with the owner, the notice to owner is not necessary. Doing so timely, should the occasion arise, gives the contractor the right to file a Mechanic's Lien for money due him on the job and subsequently to foreclose on the property.

FILING

As stated earlier, filing deadlines may vary from state to state; however, taking the State of Florida as an example, the following documents are mandatory, as of this writing, and must be filed within their designated deadlines:

- **Notice to Owner-** must be filed, by a certified service, with the owner and other parties having interest in the property, such as financial institutions, within 45 days from the day a contractor starts work or orders any material or incurs any direct expenses for the job.
- **Mechanic's Liens-** A claim of lien must be filed with the office of the clerk of the circuit court of the county in which the property is located within 90 days from the last time the contractor worked on the job.
- **Lawsuit-** A lawsuit to collect on the Mechanic's Lien must be filed within one year from the Lien's filing date.

RELEASES OF LIEN

Releases of Lien can be conditional and non-conditional, partial or full, each reflecting a specific amount and an effective period-ending date. If a contractor doesn't want to give up his rights to what is rightfully his, he must pay particular attention to the type of instrument he's signing. For example, if he receives a check in payment, he should sign a release that is conditional on the check's clearing the bank.

- **Period-ending-Date-** Regardless of which of the following releases a contractor is signing, each will show an amount and, most significantly, a period-ending date. Of the two, the period-ending date is the most important to consider. It establishes a date beyond which a contractor cannot make additional claims for money due him. In other words, a contractor states that he was paid in full up to and including that date. The amount shown on the release of lien, unless specifically provided for, has no bearing on the release of lien for that period of time. In fact, many releases, even though they are for thousands of dollars, are issued for $10 and other considerations but with a specific period-ending date.

When a contractor is most concerned with collecting his money, he may inadvertently sign a release showing the amount due him as of the 25th of the previous month with a period-ending the 30th of that month, and at times the 25th or the 30th of another month. In such a case, the contractor has released any work he has done or will do between the 26th of the previous month and the 30th of whatever that month is, even though he was not paid for it.

When signing releases of lien, always check the period-ending date as well as the amount shown.

Releases of lien come in many shapes and forms; however, the fundamental thrust of the instrument remains the same. A contractor releases his claim of lien by the amount of money or other compensations he receives for work done on a property. For every partial payment he receives toward a contract he's expected to execute a partial release. When the contractor gets the final payment, then and only then he will execute a waiver or final release of liens.

There will be times when the customer will expect the contractor to execute a partial release of lien even though he's paying him with a post-dated check. Even though it's a risky business, if a contractor decides to go along with the deal, the least he can do is execute a "conditional release of lien"—the release is valid only if the check clears the bank.

As stated earlier, lien laws, while simple to follow, vary greatly from state to state. The contractor is strongly urged to get acquainted with the laws and procedures of the state in which he intends to work.

Chapter 17

WAREHOUSING

COSTS AND BENEFITS

The warehousing of tools and material is a mainstay of most successful electrical contractors. Large contractors use warehousing to store equipment and large quantities of standard material and devices acquired at greater discount, giving them an edge on the next bid; while small contractors use warehousing to store materials left over from jobs, giving them the comfort of having parts at hand to carry out their next service call. Either way, the cost of warehousing is something that needs to be analyzed and balanced against its benefits. This task can be complex because neither cost nor benefits are tangible.

COST OF WAREHOUSING

The cost of warehousing is based on the number of square feet of usable space. For example, if the yearly rent for a 1000 sq. ft. space is $1000 then the cost per sq. ft. is $1. However, if we invest, say, in an additional 500 sq. ft. of shelving or mezzanine, then the cost per sq. ft. of usable space drops to about $0.66—a one third saving.

In laying out warehousing space, consideration has to be given to (1) the cost of operation, including rent, utilities, and maintenance personnel, (2) the costs associated with creating usable space, (3) the return on investment, and (4) the ease of operation.

Of the four elements, "ease of operation" is what we are most concerned with here, for the other costs can be easily computed and analyzed.

The financial investment a contractor makes in warehousing, regardless of its size, can only be considered cost-effective if it increases field production, if stored materials maintain at least their original value, and if the equipment depreciates at the same rate that the economics of the business and the IRS allow you to depreciate them.

For example, if material is stored in garbage-can-like containers spread throughout the floor so that the contractor has to dig his way through one half of the container just to find a fitting or two, then he should not waste this valuable space and time just for the sake of storing material. In fact, if he eliminates this kind of storage, he can reduce his warehouse size and store only equipment and tools. This creates a drastic reduction in warehousing size that can save the contractor more than it costs to buy those few fittings from a supplier when he needs them.

However, if a contractor keeps a well laid out warehouse with a shelving system where he can readily store and find what he needs, then the benefits of availability and increase in production will more than justify the additional cost of warehousing for a service oriented operation.

The cost of building a storage system that achieves the "ease of operation" goal need not be great. Normally, a system that holds ten different sizes of fittings for Flex, EMT, RGC, and PVC fittings, including a section for items such as Devices, Grounding, Wire Termination Fittings, and the like, can be assembled in a 300 sq. ft. area (see "Typical layout" at the end of this chapter). However, it could become expensive if you don't understand its principal function or fail to maintain it.

The principal function of a storage system for service work is to keep a contractor from having to run to a supplier and buy every fitting that was ever made in every size available and store it. That would be a very poor investment, for every day that that fitting is not used its cost increases in direct proportion to the contractor's cost of warehousing. Whether a contractor uses or don't use a certain fitting, rent is still paid on the space it occupies, thus taking us into that gray area of tangible and intangible warehousing costs and benefits that only the business can dictate and only the contractor can decide.

Therefore, the most economical and cost-effective storage system is the one that is built just to receive, in the least amount of space, materials left over from jobs the contractor has done. And he must store them in such a way that he can readily find them when he needs them.

A contractor should not throw away material—it's part of his last job's profit regardless of whether it came from demolition work or from the supplier. This leftover material, if well managed through the warehousing systems, besides being quickly available, can and should be turned into hard cash on the next job.

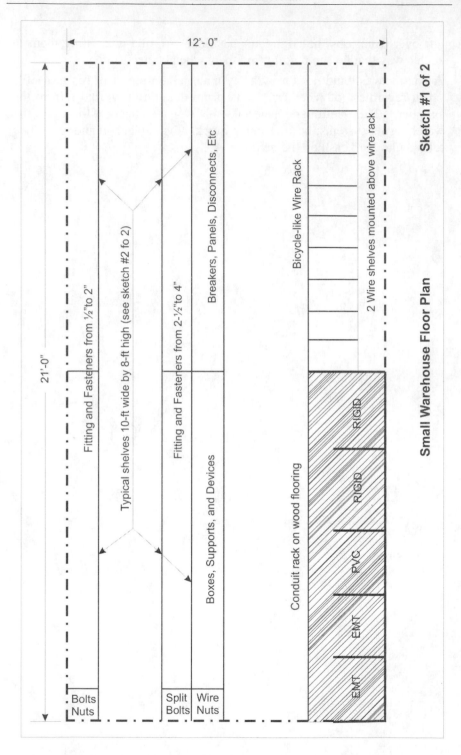

Sketch #1 of 2

12'- 0"

21'-0"

Fitting and Fasteners from ½"to 2"

Typical shelves 10-ft wide by 8-ft high (see sketch #2 fo 2)

Fitting and Fasteners from 2-½"to 4"

Breakers, Panels, Disconnects, Etc

Boxes, Supports, and Devices

Bicycle-like Wire Rack

2 Wire shelves mounted above wire rack

Conduit rack on wood flooring

RIGID

RIGID

PVC

EMT

EMT

Bolts Nuts

Split Bolts | Wire Nuts

Small Warehouse Floor Plan

A TYPICAL SHELF LAYOUT APPX. 10 ft. X 8 ft.

ASSORTED FITTINGS

EMT   LIQ. TIGHT   FLEX

1/2"   3/4"   1"   1-1/4"   1-1/2"   2"

Typical 6-1/2" W x 11"D x 4-3/4" H Bin Cartons

Regardless of the bin carton content's size, for fitting identification, always tie-wrap on the face of the carton an actual 1/2" fitting.

Sketch #2 of 2

NOTES

NOTES